Oxide Semiconductors

MATERIALS RESEARCH SOCIETY
SYMPOSIUM PROCEEDINGS VOLUME 1633

Oxide Semiconductors

Symposium held December 1–6, 2013, Boston, Massachusetts. U.S.A.

EDITORS

Steve Durbin
Western Michigan University
Kalamazoo, Michigan, U.S.A.

Marius Grundmann
Universität Leipzig
Leipzig, Germany

Anderson Janotti
University of California, Santa Barbara
Santa Barbara, California, U.S.A.

Tim Veal
University of Liverpool
Liverpool, United Kingdom

Materials Research Society
Warrendale, Pennsylvania

CAMBRIDGE UNIVERSITY PRESS
Cambridge, New York, Melbourne, Madrid, Cape Town,
Singapore, São Paulo, Delhi, Mexico City

Cambridge University Press
32 Avenue of the Americas, New York, NY 10013-2473, USA

www.cambridge.org
Information on this title: www.cambridge.org/9781605116105

Materials Research Society
506 Keystone Drive, Warrendale, PA 15086
http://www.mrs.org

© Materials Research Society 2014

This publication is in copyright. Subject to statutory exception
and to the provisions of relevant collective licensing agreements,
no reproduction of any part may take place without the written
permission of Cambridge University Press.

This book has been registered with Copyright Clearance Center, Inc.
For further information please contact the Copyright Clearance Center,
Salem, Massachusetts.

First published 2014

CODEN: MRSPDH

ISBN: 978-1-60511-610-5 Hardback

Cambridge University Press has no responsibility for the persistence or
accuracy of URLs for external or third-party Internet Web sites referred to
in this publication and does not guarantee that any content on such Web sites is,
or will remain, accurate or appropriate.

CONTENTS

Preface ix

Materials Research Society Symposium Proceedings xi

SYNTHESIS

*Synthesis and Characterization of Copper Oxide Compounds 3
K.P. Hering, A. Polity, B. Kramm,
A. Portz, and B.K. Meyer

Characterization of Tin Oxide Grown by Molecular
Beam Epitaxy 13
G. Medina, P.A. Stampe, R.J. Kennedy, R.J. Reeves,
G.T. Dang, A. Hyland, M.W. Allen,
M.J. Wahila, L.F.J. Piper, and S.M. Durbin

Epitaxial Growth of $(Na,K)NbO_3$ Based Materials on $SrTiO_3$
by Pulsed Laser Deposition 19
T. Hanawa, N. Kikuchi, K. Nishio, K. Tonooka,
R. Wang, and T. Mamiya

Structural and Electrical Properties of $LaNiO_3$ Thin Films Grown
on (100) and (001) Oriented $SrLaAlO_4$ Substrates by
Chemical Solution Deposition Method 25
D.S.L. Pontes, F.M Pontes, Marcelo A. Pereira-da-Silva,
O.M. Berengue, A.J. Chiquito, and E. Longo

OPTICAL AND ELECTRICAL CHARACTERIZATION

Native Point Defects in Multicomponent Transparent
Conducting Oxides 37
Altynbek Murat and Julia E. Medvedeva

*Invited Paper

Electronic Transport Characterization of BiVO$_4$ Using AC Field Hall Technique 43
Jeffery Lindemuth, Alexander J.E. Rettie,
Luke G. Marshall, Jianshi Zhou, and C. Buddie Mullins

A DLTS Study of a ZnO Microwire, a Thin Film and Bulk Material..51
Florian Schmidt, Peter Schlupp, Stefan Müller,
Christof Peter Dietrich, Holger von Wenckstern,
Marius Grundmann, Robert Heinhold,
Hyung-Suk Kim, and Martin Ward Allen

Evaluation of Sub-gap States in Amorphous In-Ga-Zn-O Thin Films Treated with Various Process Conditions 55
Kazushi Hayashi, Aya Hino, Hiroaki Tao,
Yasuyuki Takanashi, Shinya Morita, Hiroshi Goto,
and Toshihiro Kugimiya

Effects of N$_2$O Addition on the Properties of ZnO Thin Films Grown Using High-temperature H$_2$O Generated by Catalytic Reaction 61
Naoya Yamaguchi, Eichi Nagatomi,
Takahiro Kato, Koichiro Ohishi,
Yasuhiro Tamayama, and Kanji Yasui

Density Functional Study of Benzoic Acid Derivatives Modified SnO$_2$ (110) Surface 69
Tegshjargal Khishigjargal and Kazuyoshi Ueda

Defect Driven Emission from ZnO Nano Rods Synthesized by Fast Microwave Irradiation Method for Optoelectronic Applications....................................75
Nagendra Pratap Singh, S.A. Shivashankar,
and Rudra Pratap

Breaking of Raman Selection Rules in Cu$_2$O by Intrinsic Point Defects..81
Thomas Sander, Christian T. Reindl, and Peter J. Klar

Characterization of Mechanical, Optical and Structural Properties of Bismuth Oxide Thin Films as a Write-once Medium for Blue Laser Recording............................87
M. Martyniuk, D. Baldwin, R. Jeffery, K.K.M.B.D. Silva,
R.C. Woodward, J. Cliff, R.N. Krishnan,
J.M. Dell, and L. Faraone

DEVICE ISSUES

High Performance IGZO TFTs with Modified Etch Stop Structure on Glass Substrates 95
  Forough Mahmoudabadi, Ta-Ko Chuang,
  Jerry Ho Kung, and Miltiadis K. Hatalis

Amorphous Zinc-tin Oxide Thin Films Fabricated by Pulsed Laser Deposition at Room Temperature 101
  P. Schlupp, H. von Wenckstern, and M. Grundmann

Solution Processed Resistive Random Access Memory Devices for Transparent Solid-state Circuit Systems 105
  Yiran Wang, Bing Chen, Dong Liu, Bin Gao, Lifeng Liu,
  Xiaoyan Liu, and Jinfeng Kang

Structural and Electrical Characteristics of Ternary Oxide $SmGdO_3$ for Logic and Memory Devices 111
  Yogesh Sharma, Pankaj Misra, and Ram S. Katiyar

Correlation of Resistance Switching Behaviors with Dielectric Functions of Manganite Films: A Study by Spectroscopic Ellipsometry 117
  Masaki Yamada, Toshihiro Nakamura, and Osamu Sakai

A Continuous Composition Spread Approach Towards Monolithic, Wavelength-selective Multichannel UV-photo-detector Arrays 123
  H. von Wenckstern, Z. Zhang, J. Lenzner,
  F. Schmidt, and M. Grundmann

Metal-semiconductor-insulator-metal Structure Field-effect Transistors Based on Zinc Oxides and Doped Ferroelectric Thin Films 131
  Ze Jia, Jianlong Xu, Xiao Wu, Mingming Zhang,
  Naiwen Zhang, Jizhi Liu, Zhiwei Liu, and Juin J. Liou

Highly Reliable Passivation Layer for a-InGaZnO Thin-film Transistors Fabricated Using Polysilsesquioxane 139
  Juan Paolo Bermundo, Yasuaki Ishikawa,
  Haruka Yamazaki, Toshiaki Nonaka, and Yukiharu Uraoka

Author Index 145

Subject Index 147

PREFACE

Symposium R, "Oxide Semiconductors" was held Dec. 1–Dec. 6 at the 2013 MRS Fall Meeting in Boston, Massachusetts.

Oxide semiconductors are poised to take a more active role in modern electronics, particularly in the field of thin film transistors. While many advances have been made in terms of our understanding of fundamental optical and electronic characteristics, there remain many questions in terms of defects, doping and optimal growth/synthesis conditions.

This symposium proceedings volume represents recent advances in growth and characterization of a number of different oxide semiconductors, as well as device fabrication. The papers are divided into three sections: (1) Synthesis (2) Optical and Electrical Characterization and (3) Device Issues. In selecting these papers, it is our hope that readers get a sense of the many interesting discussions that were held during the symposium, and the excitement in the community about this class of materials. We would also like to take this opportunity to thank the Army Research Office for financial support. The views, opinions, and/or findings contained in this report are those of the author(s) and should not be construed as an official Department of the Army position, policy, or decision, unless so designated by other documentation.

Steve Durbin
Marius Grundmann
Anderson Janotti
Tim Veal

March 2014

MATERIALS RESEARCH SOCIETY SYMPOSIUM PROCEEDINGS

Volume 1607E — Nanotechnology-Enhanced Coatings, 2014, A. Taylor, N. Ludford, C. Avila-Orta, C. Becker-Willinger, ISBN 978-1-60511-584-9
Volume 1608 — Nanostructured Materials and Nanotechnology, 2014, J.L. Rodríguez-López, O. Graeve, A.G. Palestino-Escobedo, M. Muñoz-Navia, ISBN 978-1-60511-585-6
Volume 1609E — Biomaterials for Medical Applications, 2014, S. E. Rodil, A. Almaguer-Flores, K. Anselme, J. Castro, ISBN 978-1-60511-586-3
Volume 1610E — Advanced Materials and Technologies for Energy Storage Devices, 2014, I. Belharouak, J. Xiao, P. Balaya, D. Carlier, A. Cuentas-Gallegos, ISBN 978-1-60511-587-0
Volume 1611 — Advanced Structural Materials, 2014, H. Calderon, H. Balmori Ramirez, A. Salinas Rodriguez, ISBN 978-1-60511-588-7
Volume 1612E — Concrete and Durability of Concrete Structures, 2013, L. E. Rendon Diaz Miron, B. Martínez Sánchez, N. Ramirez Salinas, ISBN 978-1-60511-589-4
Volume 1613 — New Trends in Polymer Chemistry and Characterization, 2014, L. Fomina, G. Cedillo Valverde, M. del Pilar Carreón Castro, ISBN 978-1-60511-590-0
Volume 1614E — Advances in Computational Materials Science, 2014, E. Martínez Guerra, J.U. Reveles, O. de la Peña Seaman, ISBN 978-1-60511-591-7
Volume 1615E — Electron Microscopy of Materials, 2014, H. Calderon, C. Kisielowski, L. Francis, P. Ferreira, A. Mayoral, ISBN 978-1-60511-592-4
Volume 1616 — Structural and Chemical Characterization of Metals, Alloys and Compounds, 2014, A. Contreras Cuevas, R. Pérez Campos, R. Esparza Muñoz, ISBN 978-1-60511-593-1
Volume 1617 — Low-Dimensional Semiconductor Structures, 2013, T.V. Torchynska, L. Khomenkova, G. Polupan, G. Burlak, ISBN 978-1-60511-594-8
Volume 1618 — Cultural Heritage and Archaeological Issues in Materials Science (CHARIMSc), 2014, J.L. Ruvalcaba Sil, J. Reyes Trujeque, A. Velázquez Castro, M. Espinosa Pesqueira, ISBN 978-1-60511-595-5
Volume 1619E — Modeling and Theory-Driven Design of Soft Materials, 2014, M. Rodger, M. Dutt, Y. Yingling, V. Ginzburg, ISBN 978-1-60511-596-2
Volume 1621 — Advances in Structures, Properties and Applications of Biological and Bioinspired Materials, 2014, T. Deng, H. Liu, S.P. Nukavarapu, M. Oyen, C. Tamerler, ISBN 978-1-60511-598-6
Volume 1622 — Fundamentals of Gels and Self-Assembled Polymer Systems, 2014, F. Horkay, N. Langrana, M. Shibayama, S. Basu, ISBN 978-1-60511-599-3
Volume 1623E — Synthetic Tools for Understanding Biological Phenomena, 2014, D. Benoit, A. Kloxin, C-C. Lin, V. Sée, ISBN 978-1-60511-600-6
Volume 1624E — Integration of Biomaterials with Organic Electronics, 2014, M.R. Abidian, M. Irimia-Vladu, R. Owens, M. Rolandi, ISBN 978-1-60511-601-3
Volume 1625E — Multiscale Materials in the Study and Treatment of Cancer, 2014, N. Moore, S. Peyton, J. Snedeker, C. Williams, ISBN 978-1-60511-602-0
Volume 1626 — Micro- and Nanoscale Processing of Materials for Biomedical Devices, 2014, R. Narayan, V. Davé, S. Jayasinghe, M. Reiterer, ISBN 978-1-60511-603-7
Volume 1627E — Photonic and Plasmonic Materials for Enhanced Optoelectronic Performance, 2014, C. Battaglia, ISBN 978-1-60511-604-4
Volume 1628E — Large-Area Processing and Patterning for Active Optical and Electronic Devices, 2014, T.D. Anthopoulos, I. Kymissis, B. O'Connor, M. Panzer, ISBN 978-1-60511-605-1
Volume 1629E — Functional Aspects of Luminescent and Photoactive Organic and Soft Materials, 2014, M.C. Gather, S. Reineke, ISBN 978-1-60511-606-8
Volume 1630E — Solution Processing of Inorganic and Hybrid Materials for Electronics and Photonics, 2014, P. Smith, M. van Hest, H. Hillhouse, ISBN 978-1-60511-607-5
Volume 1631E — Emergent Electron Transport Properties at Complex Oxide Interfaces, 2014, K.H. Bevan, S. Ismail-Beigi, T.Z. Ward, Z. Zhang, ISBN 978-1-60511-608-2
Volume 1632E — Organic Microlasers—From Fundamentals to Device Application, 2014, R. Brückner, ISBN 978-1-60511-609-9
Volume 1633 — Oxide Semiconductors, 2014, S. Durbin, M. Grundmann, A. Janotti, T. Veal, ISBN 978-1-60511-610-5
Volume 1634E — Diamond Electronics and Biotechnology—Fundamentals to Applications VII, 2014, J. C. Arnault, C.L. Cheng, M. Nesladek, G.M. Swain, O.A. Williams, ISBN 978-1-60511-611-2

MATERIALS RESEARCH SOCIETY SYMPOSIUM PROCEEDINGS

Volume 1635 —	Compound Semiconductor Materials and Devices, 2014, F. Shahedipour-Sandvik, L.D. Bell, K.A. Jones, A. Clark, K. Ohmori, ISBN 978-1-60511-612-9
Volume 1636E —	Magnetic Nanostructures and Spin-Electron-Lattice Phenomena in Functional Materials, 2014, A. Petford-Long, ISBN 978-1-60511-613-6
Volume 1638E —	Next-Generation Inorganic Thin-Film Photovoltaics, 2014, C. Kim, C. Giebink, B. Rand, A. Boukai, ISBN 978-1-60511-615-0
Volume 1639E —	Physics of Organic and Hybrid Organic-Inorganic Solar Cells, 2014, P. Ho, M. Niggemann, G. Rumbles, L. Schmidt-Mende, C. Silva, ISBN 978-1-60511-616-7
Volume 1640E —	Sustainable Solar-Energy Conversion Using Earth-Abundant Materials, 2014, S. Jin, K. Sivula, J. Stevens, G. Zheng, ISBN 978-1-60511-617-4
Volume 1641E —	Catalytic Nanomaterials for Energy and Environment, 2014, J. Erlebacher, D. Jiang, V. Stamenkovic, S. Sun, J. Waldecker, ISBN 978-1-60511-618-1
Volume 1642E —	Thermoelectric Materials—From Basic Science to Applications, 2014, Q. Li, W. Zhang, I. Terasaki, A. Maignan, ISBN 978-1-60511-619-8
Volume 1643E —	Advanced Materials for Rechargeable Batteries, 2014, T. Aselage, J. Cho, B. Deveney, K.S. Jones, A. Manthiram, C. Wang, ISBN 978-1-60511-620-4
Volume 1644E —	Materials and Technologies for Grid-Scale Energy Storage, 2014, B. Chalamala, J. Lemmon, V. Subramanian, Z. Wen, ISBN 978-1-60511-621-1
Volume 1645 —	Advanced Materials in Extreme Environments, 2014, M. Bertolus, H.M. Chichester, P. Edmondson, F. Gao, M. Posselt, C. Stanek, P. Trocellier, X. Zhang, ISBN 978-1-60511-622-8
Volume 1646E —	Characterization of Energy Materials *In-Situ* and *Operando*, 2014, I. Arslan, Y. Gogotsi, L. Mai, E. Stach, ISBN 978-1-60511-623-5
Volume 1647E —	Surface/Interface Characterization and Renewable Energy, 2014, R. Opila, F. Rosei, P. Sheldon, ISBN 978-1-60511-624-2
Volume 1648E —	Functional Surfaces/Interfaces for Controlling Wetting and Adhesion, 2014, D. Beysens, ISBN 978-1-60511-625-9
Volume 1649E —	Bulk Metallic Glasses, 2014, S. Mukherjee, ISBN 978-1-60511-626-6
Volume 1650E —	Materials Fundamentals of Fatigue and Fracture, 2014, A.A. Benzerga, E.P. Busso, D.L. McDowell, T. Pardoen, ISBN 978-1-60511-627-3
Volume 1651E —	Dislocation Plasticity, 2014, J. El-Awady, T. Hochrainer, G. Po, S. Sandfeld, ISBN 978-1-60511-628-0
Volume 1652E —	Advances in Scanning Probe Microscopy, 2014, T. Mueller, ISBN 978-1-60511-629-7
Volume 1653E —	Neutron Scattering Studies of Advanced Materials, 2014, J. Lynn, ISBN 978-1-60511-630-3
Volume 1654E —	Strategies and Techniques to Accelerate Inorganic Materials Innovation, 2014, S. Curtarolo, J. Hattrick-Simpers, J. Perkins, I. Tanaka, ISBN 978-1-60511-631-0
Volume 1655E —	Solid-State Chemistry of Inorganic Materials, 2014, S. Banerjee, M.C. Beard, C. Felser, A. Prieto, ISBN 978-1-60511-632-7
Volume 1656 —	Materials Issues in Art and Archaeology X, 2014, P. Vandiver, W. Li, C. Maines, P. Sciau, ISBN 978-1-60511-633-4
Volume 1657E —	Advances in Materials Science and Engineering Education and Outreach, 2014, P. Dickrell, K. Dilley, N. Rutter, C. Stone, ISBN 978-1-60511-634-1
Volume 1658E —	Large-Area Graphene and Other 2D-Layered Materials—Synthesis, Properties and Applications, 2014, editor TBD, ISBN 978-1-60511-635-8
Volume 1659 —	Micro- and Nanoscale Systems – Novel Materials, Structures and Devices, 2014, J.J. Boeckl, R.N. Candler, F.W. DelRio, A. Fontcuberta i Morral, C. Jagadish, C. Keimel, H. Silva, T. Voss, Q. H. Xiong, ISBN 978-1-60511-636-5
Volume 1660E —	Transport Properties in Nanocomposites, 2014, H. Garmestani, H. Ardebili, ISBN 978-1-60511-637-2
Volume 1661E —	Phonon-Interaction-Based Materials Design—Theory, Experiments and Applications, 2014, D.H. Hurley, S.L. Shinde, G.P. Srivastava, M. Yamaguchi, ISBN 978-1-60511-638-9
Volume 1662E —	Designed Cellular Materials—Synthesis, Modeling, Analysis and Applications, 2014, K. Bertoldi, ISBN 978-1-60511-639-6
Volume 1663E —	Self-Organization and Nanoscale Pattern Formation, 2014, M.P. Brenner, P. Bellon, F. Frost, S. Glotzer, ISBN 978-1-60511-640-2

Volume 1664E — Elastic Strain Engineering for Unprecedented Materials Properties, 2014, J. Li, E. Ma, Z. W. Shan, O.L. Warren, ISBN 978-1-60511-641-9

Prior Materials Research Symposium Proceedings available by contacting Materials Research Society

Synthesis

Synthesis and Characterization of Copper Oxide Compounds

K.P. Hering, A. Polity, B. Kramm, A. Portz and B.K. Meyer

I. Physics Institute, Justus-Liebig-University Giessen, Heinrich-Buff-Ring 16, 35392 Giessen, Germany

ABSTRACT

The p-type conducting Copper-oxide compound semiconductors (Cu_2O, CuO) provide a unique possibility to tune the band gap energies from 2.1 eV to the infrared at 1.40 eV into the middle of the efficiency maximum for solar cell applications. By a pronounced non-stoichiometry the electronic properties may vary from insulating to metallic conduction. They appear to be an attractive alternative absorber material in terms of abundance, sustainability, non-toxicity of the elements, and numerous methods for thin film deposition that facilitate low cost production. The synthesis and characterization of Cu_2O thin films used as p-type absorbers in heterojunction solar cells will be reported. We discuss properties of the undoped non-stoichiometric Cu_2O, controlled p-type doping by nitrogen, analysis of band offsets by X-ray photoelectron spectroscopy (XPS). In addition we show proof of concept for an increase in photovoltaic conversion efficiency in AlGaN/Cu_2O heterostructures due to a more favorable band alignment.

INTRODUCTION

AlGaAs/GaAs ternary III-V compound semiconductors revolutionized the field of semiconductor heterostructures and opened the door towards high speed low dimensional devices. The approach was to synthesize by epitaxial growth techniques well-ordered monocrystalline layers with abrupt and atomically sharp interfaces by combining two semiconductors with different band gaps, effective masses etc. while retaining the lattice matching between the two compounds as good as possible. This concept has been widely used in the last thirty years for almost all III-V and II-VI semiconductors and their ternary and quaternary alloys with a wide range of applications from optoelectronics to solar cells.

A completely different and new concept can be visualized and realized in pure binary oxide compounds such as the copper oxides or tin oxides. Here at a fixed metal to oxygen ratio one stabilizes distinct different phases such as Cu_2O, Cu_4O_3 and CuO which have unique structural (crystal symmetry from cubic, tetragonal to monoclinic), optical (band gap energies from 2.1 to 1.4 eV, direct and indirect band gaps) and electronic (mobilities, carrier concentrations, band offsets) properties. A defined non-stoichiometry exists for all stable phases, which allows tuning the defect disorder and related to it the electronic properties from insulating to metallic conduction. The copper compounds are intrinsically p-conducting oxides and bi-polar doping is still at an exploratory phase. Surface conductivity is different from bulk conductivity as is ionic conduction from electronic conduction. For many applications (gas sensors, heterojunction solar cells and thin film transistors) polycrystalline materials are used and the routes towards high ordered epitaxial growth giving bench marks for the materials' quality have not been investigated. [1-10]

Identifying new, prospective p-type absorber layers for heterojunction solar cells has led to the CuInGa(Se)$_2$ system, and by taking into account sustainability and scarcity of elements to the Kesterites. The copper oxides are an attractive alternative in terms of availability of components, sustainability, a large variety of thin film synthesis methods offering low cost production, and non-toxicity of the elements. For the future of a copper oxide based thin film solar cell systems (including tandem arrangement by two different copper oxide phases) physical properties have to be investigated, established, improved and tested in full devices. [11-14]

EXPERIMENT

Cu$_2$O thin films studied in this work were deposited by radio-frequency (RF) magnetron sputtering. A high-purity (99.999%) 4" ceramic Cu$_2$O or metallic Cu target was used. As sputtering gas, argon or a mixture of argon, oxygen and nitrogen have been used. The RF-power was mostly fixed at 300 W. During deposition, the working pressure in the sputtering chamber was usually kept between 0.7 and 2 Pa. The Cu$_2$O films were grown on glass substrates, sapphire or GaN/AlGaN-templates.
The crystallinity was investigated with X-ray diffraction (XRD) using a SIEMENS D5000 diffractometer with a Cu-K$_\alpha$ X-ray source operating in Bragg-Brentano geometry. As-grown, undoped and nitrogen doped films showed a preferential (200) orientation. Surface morphologies and grain sizes were examined by an atomic force (AFM) microscope (Quesant Q-250) operating in noncontact mode. For low nitrogen contents the grain sizes were between 50 to 60 nm and increased to 200-300 nm for the highest nitrogen flows used. Surface roughness was between 4.2 and 5.5 nm independent of doping level. The amount of nitrogen in the films was determined by secondary ion mass spectrometry (CAMECA Riber). In order to be quantitative the sensitivity factor of nitrogen in Cu$_2$O was evaluated using implantation standards. For the evaluation of the band offsets thin films of the different copper oxides (thicknesses of the order of 40 nm) were deposited on GaN-templates. Subsequent Argon sputtering to remove the material is used to allow for XPS measurements (PHI Electronics VersaProbe II), in the copper oxide films, in the interface region between the copper oxides and GaN and in the GaN layer. For device characterization, J-V characteristics and EQE were measured with a Keithley 4200SCS parameter analyzer, combined with an Oriel Xe arc lamp, Oriel AM1.5g and a monochromator.

DISCUSSION

Cu$_2$O is one of the rare intrinsically p-conducting semiconductors, and in addition has its energy band gap in the visible spectral range. A stick and ball model of the crystal structure is shown in Fig.1. The Cu$_2$O lattice has cuprite cubic structure; each copper atom is linearly coordinated by two oxygen atoms. The Cu$_2$O (001) surface is polar and can be oxygen or copper terminated. The Cu$_2$O (111) surface is non polar.

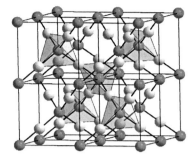

Fig.1: Stick and ball model of the crystal structure of Cu_2O (grey: Cu^+, red: O^{2-}; from wikipedia.org)

Cu_2O is considered to be a promising material for thin-film-solar cell applications. Sputtering from metallic and ceramic targets can conveniently be used for the thin-film deposition of copper oxide. Depending on the oxygen partial-pressure, the properties of Cu_xO are adjustable over a wide range of the binary compounds, Cu_2O and CuO, respectively. The electrical properties change considerably depending on the copper-to-oxygen-ratio. It is commonly assumed that copper-vacancies and related defects are the dominant intrinsic acceptors. The free hole carrier concentrations in Cu_2O can vary between 10^{14} to 10^{20} cm^{-3} in dependence of the non-stoichiometry (oxygen flow, see Fig. 2).

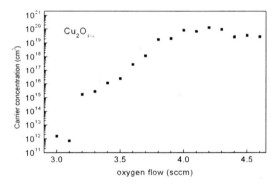

Fig.2: Free hole carrier concentration of Cu_2O films prepared under different oxygen flows (metallic target). Deposition was on unheated quartz glass substrates.

For many applications it is appropriate to control and to fine tune the carrier concentration especially for p-i-n heterostructure diodes where the p and i side of the junction are made from Cu_2O.

There are only a few reports on the controlled doping of Cu_2O with acceptors and one of the successful candidates seems to be nitrogen [15]. We therefore investigated the properties of Cu_2O-thin-films with additional nitrogen as a reactive gas in a RF magnetron sputter process. We quantify the amount of nitrogen by secondary ion mass spectrometry and determine the electrical properties (resistivity, mobility and carrier concentration) from temperature dependent Hall effect measurements.

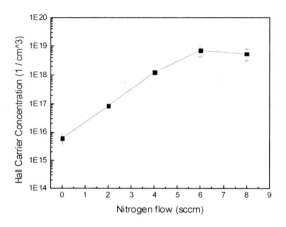

Fig.3: Free hole carrier concentrations as determined by Hall-effect measurements at room temperature as a function of the nitrogen flow in RF-sputtered Cu_2O films.

As can be seen in Fig.3 the total amount of charge carriers increases as a function of the nitrogen flow from 5×10^{15} cm^{-3} (undoped prepared with an oxygen flow of 3.2 sccm according to Fig.2) to close to 10^{19} cm^{-3}. The hole concentrations can thus be adopted to the number of electron hole pairs created in the absorption process under AM1.5 illumination and in the highly doped regime acting as back surface field and minimizing contact resistance.

Most of the solar cell prototypes tested so far worked with p-Cu_2O as absorber and n-ZnO as window material. One crucial point in those device structures was that the valence and conduction band discontinuities and the actual band offsets between ZnO and Cu_2O have not been taken into account for an explanation of the low efficiencies of the photovoltaic devices. The valence band structures of Cu_2O and Cu_4O_3 were studied by X-ray photoelectron spectroscopy (XPS). Additionally we investigated ZnO(GaN)/Cu_2O and GaN/Cu_4O_3 layer structures to get insight into the alignment of the bands. By XPS for ZnO/Cu_2O the band offsets were determined to be 2.17 eV in the valence band and 0.97 eV in the conduction band [20]. An alternative window material is GaN, and from the XPS investigations the conduction band offset was determined to be at best 0.9 eV, see Fig.4.

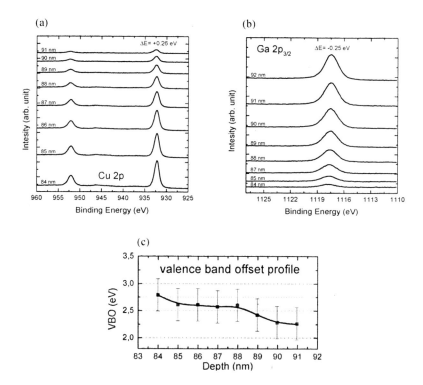

Fig.4: Section of the XPS spectrum showing (a) the Cu $2p_{1/2}$ and Cu $2p_{3/2}$ signals of the Cu_2O film and (b) the Ga $2p_{3/2}$ signal of the GaN film by step-by-step removal of the respective materials by Argon sputtering. The evolution of the corresponding calculated valence band offsets throughout the interface region is depicted in (c).

However, aligning the conduction bands to completely avoid a conduction band offset in order to prevent the corresponding conversion efficiency losses in a heterojunction solar cell can be made by alloying GaN with Al. A typical device structure of a heterojunction solar cell starts from the glass substrate, followed by a highly n-type (metallic) transparent conducting material (GaN, ZnO, MgZnO). The window layer can consist of the same materials but at lower doping levels. A low doped p-absorber and a highly doped p-layer complete the structure. Finally metallic contacts are placed.

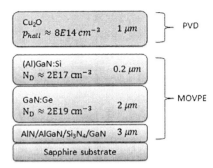

Fig.5: Schematic photovoltaic device structure of the produced photovoltaic cells.

In order to prove the concept of a gain in conversion efficiency, lowering the conduction band offset by alloying GaN with Al, experiments were carried out, employing RF-sputtered cuprous oxide thin films in conjunction with AlGaN templates. The templates were grown via MOVPE. The cuprous oxide thin films were produced via RF sputtering, where 50 sccm Ar and 3.4 sccm O_2, corresponding to 0.6 Pa working pressure, were used at a substrate temperature of 900 K to ensure acceptable electronic properties. The cuprous oxide thin films were grown on all three templates, different in aluminum concentration, in the same deposition. After deposition, to complete the photovoltaic cell, photolithographic processing and chemical etching steps were undertaken, followed by the deposition of Ti/Au and Au metal contacts via thermal evaporation for AlGaN and Cu_2O, respectively. The J-V characteristics were carried out under AM1.5g-illumination at 100 mW/cm².

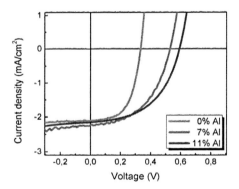

Fig.6: Illuminated J-V characteristics of the heterostructures employing different aluminum concentrations in the AlGaN templates used.

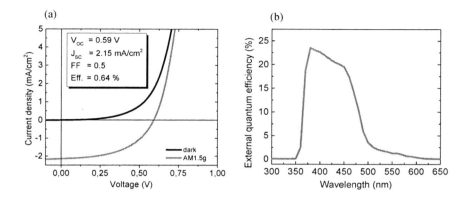

Fig.7: (a) Dark and illuminated J-V characteristics (b) external quantum efficiency of the cell using highest aluminum concentration (11%) in its template.

While the short circuit current densities stay almost constant, the open circuit voltage scales with increased aluminum content in the window layer, see Fig.6, resulting in an open circuit voltage of 0.59 V for the highest aluminum concentration used. The poor fill factors can be attributed to parasitic resistance, i.e. series resistance. The external quantum efficiency decreases with increasing wavelength, showing a strong decline at the first direct allowed transition of the cuprous oxide absorber, clearly indicating a low minority carrier lifetime. The overall low quantum efficiency can be explained by a strong recombination path at the heterointerface, most likely resulting from the sputtering damage caused during the deposition. See Fig.7.

Nothing is known so far about the band offsets within the copper oxide compounds, e.g., Cu_2O with respect to Cu_4O_3 or CuO. The knowledge about the band offsets between the three different copper compounds and the band alignment of the bands in heterostructures with ZnO and GaN will have an influence on the realization of photovoltaic devices and applications in single and tandem solar cells. We therefore started the investigation of Cu_4O_3 on GaN and report here on the actual data.

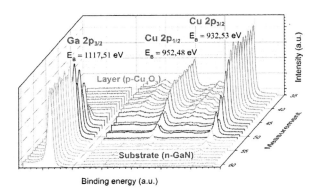

Fig.8: Section of the XPS spectrum showing the Cu $2p_{1/2}$ and Cu $2p_{3/2}$ signals of the Cu_4O_3 film and the Ga $2p_{3/2}$ signal of the GaN film by a step-by-step removal of the respective materials by Argon sputtering.

After deposition of Cu_4O_3 on GaN the surface by soft Ar sputtering was cleaned to remove carbon contaminations. As soon as the Carbon signal disappeared a first spectrum was taken. Then part of the layer was removed by Ar-sputtering and the next spectrum was taken. This procedure was continued until only signals of the GaN-template were detectable. Special care was taken to adjust the sputter rates in order to have full information on the interface region and minimize changes in stoichiometry due to preferential sputtering.

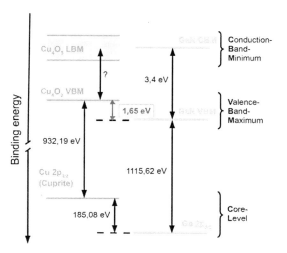

Fig.9: Binding energies of the core levels with respect to the valence band maxima in Cu_4O_3, in GaN and in the interface. The valence band offset was determined to 1.65 eV.

With the information on the binding energies of the core levels with respect to the valence band maxima in Cu_4O_3, in GaN and in the interface the valence band offset was determined to 1.65 eV. The conduction band offset could not be determined since it requires a precise knowledge of the band gap energy (see Fig.9). Temperature dependent optical measurements are currently underway to establish this value.

RESULTS

In dependence of the non-stoichiometry, the free hole carrier concentrations in Cu_2O can vary between 10^{14} to 10^{20} cm^{-3}. In Cu_2O, prepared with an oxygen flow of 3.2 sccm, the total amount of free charge carriers increases as a function of the added nitrogen flow from 5×10^{15} cm^{-3} to close to 10^{19} cm^{-3}. Both, controlled non-stoichiometry and addition of nitrogen, can therefore be applied to create a highly doped regime acting as back surface field and minimizing contact resistance in a photovoltaic device. The valence band offsets of GaN/Cu_2O and GaN/Cu_4O_3 have been determined to 2.5 eV and 1.65 eV, respectively. $AlGaN/Cu_2O$ heterostructures were processed into photovoltaic cells, showing an increase in open circuit voltage, and consequently a conversion efficienct, depending on the aluminum concentration used. The limiting influence of the conduction band offset in these heterostructures is clearly visible, suggesting, given a sufficient dupability, even higher aluminum concentrations.

ACKNOWLEDGMENTS

We would like to thank Alois Krost and Armin Dadgar, Otto-von-Guericke Universität Magdeburg for providing the AlGaN templates.

REFERENCES

[1] C. Malerba, F. Biccari, C. Leonor, A. Ricardo, M. D'Incau, P. Scardi, and A. Mittiga, Sol. Energ. Mater. Sol. C. **95**, 2848 (2011).
[2] B. K. Meyer, A. Polity, D. Reppin, M. Becker, P. Hering, P.J. Klar, Th. Sander, C. Reindl, J. Benz, M. Eickhoff, C. Heiliger, M. Heinemann, J. Bläsing, A. Krost, and S. Shokovets, phys. stat. sol. (b) **249**, 1487 (2012).
[3] L. O. Grondahl, Phys. Rev. **27**, 813 (1926).
[4] L. O. Grondahl, U. S. Pat. No. 1,640, 335 (1927).
[5] L.O. Grondahl, and P.H. Geiger, J. Am. Inst. Elec. Eng. **46**, 215 (1927).
[6] A. de Vos, J. Phys. D: Appl. Phys. **13**, 839 (1980).
[7] B. K. Meyer, S. Merita, and A. Polity, phys. stat. sol. RRL **7**, 360 (2013).
[8] Y. Ching, Yong-Nian Xu, and K. W. Wong, Phys. Rev. B **40**, 7684 (1989).
[9] D. Wu, Q. Zhang, and M. Tao, Phys. Rev. B **73**, 235206 (2006).
[10] J. Ghijsen, L. H. Tjeng, J. van Elp, H. Eskes, J. Westerink, G. A. Sawatzky, and M. T. Czyzyk, Phys. Rev. B **38**, 11322 (1988).
[11] Global Resource Depletion, Managed Austerity and the Elements of Hope, A. Diederen, Eburon Academic Publishers (2010)
[12] http://ec.europa.eu/enterprise/policies/rawmaterials/documents/index_en.htm
[13] S. Siebentritt and S. Schnorr, Progress in Photovoltaics **20**, 512 (2012).
[14] Earth-Abundant Thin-Film Photovoltaics, Workshop 28./29.3.2013, Caltech, CA, USA
[15] K. Akimoto, S. Ishizuka, M. Yanagita, Y. Nawa, G. K. Paul, T. Sakurai, Solar Energy **80**, 715 (2006).
[16] J. Herion, E. Niekisch, G. Scharl, Solar Energy Materials **4**, 101 (1980).
[17] A. Mittiga, E. Salza, F. Sarto, M. Tucci, R. Vasanthi, Appl. Phys. Lett. **88**, 163502 (2006)
[18] Y. Nishi, T. Miyata, and T. Minami, Thin Solid Films **528**, 72 (2013)
[19] T. Minami, Y. Nishi, and T. Miyata, Applied Physics Express **6**, 044101 (2013)
[20] B. Kramm, A. Laufer, D. Reppin, A. Kronenberger, P. Hering, A. Polity, and B. K. Meyer, Appl. Phys. Lett. **100**, 094102 (2012).

Mater. Res. Soc. Symp. Proc. Vol. 1633 © 2014 Materials Research Society
DOI: 10.1557/opl.2014.305

Characterization of Tin Oxide Grown by Molecular Beam Epitaxy

G. Medina[1], P.A. Stampe[2], R.J. Kennedy[2], R.J. Reeves[3], G.T. Dang[4], A. Hyland[4], M.W. Allen[4], M.J. Wahila[5], L.F.J. Piper[5], and S. M. Durbin[1,6,7]

[1]Department of Electrical Engineering, University at Buffalo, Buffalo, NY 14260, USA
[2]Department of Physics, Florida A&M University, Tallahassee, FL 32307, USA
[3]Department of Physics, University of Canterbury, Christchurch 8140, New Zealand
[4]Department of Electrical and Computer Engineering, University of Canterbury, Christchurch 8140, New Zealand
[5]Department of Physics, Binghamton University, Binghamton, NY 13902, USA
[6]Department of Physics, University at Buffalo, Buffalo, NY 14260, USA
[7]Department of Electrical and Computer Engineering, Western Michigan University, Kalamazoo, MI 49008, USA

ABSTRACT

We describe the characteristics of a series of thin film tin oxide films grown by plasma-assisted molecular beam epitaxy on r-plane sapphire substrates over a range of flux and substrate temperature conditions. A mixture of both SnO_2 and SnO are detected in several films, with the amount depending on growth conditions, most particularly the substrate temperature. Electrical measurements were not possible on all samples due to roughness related issues with contacting, but at least one film exhibited p-type characteristics depending on measurement conditions, and one sample exhibited significant persistent photoconductivity upon ultraviolet excitation in a metal-semiconductor-metal device structure.

INTRODUCTION

Tin oxide is a particularly interesting semiconductor, in that two different compounds are readily grown in thin film form: SnO_2, with a band gap energy of approximately 3.5 eV, and metastable SnO, with a somewhat narrower, less well characterized band gap energy. SnO_2 has many similarities with ZnO, another ultraviolet band gap semiconductor, including a strong tendency to be n-type as grown, as well as a surface electron accumulation layer [1,2]. Both of these characteristics are related to a charge neutrality level near the conduction band edge, and together can interfere with both pn junction formation and Schottky based device fabrication. In contrast, the narrower band gap compound SnO is intrinsically p-type. Due to the metastable nature of this material it must be grown at lower temperature, above which mixed phase or purely SnO_2 may result.
SnO_2 is easily grown by a variety of techniques including RF sputtering [3], pulsed laser deposition [4], and chemical vapor deposition [5]. There are fewer reports of SnO growth, although it can be achieved by careful control of deposition conditions [6]. SnO_2 has also been grown by molecular beam epitaxy (MBE) [7-9]. With the goal of achieving predominantly SnO single crystal films with an epitaxial relationship with the substrate for subsequent device applications, we report in this work preliminary results on the growth by plasma assisted MBE (PA-MBE) and characterization of tin oxide films.

EXPERIMENT

Films were grown in a custom-built, user-modified SVTA molecular beam epitaxy system having a base pressure of approximately 5×10^{-10} torr. Tin was evaporated in a standard effusion cell, and flux was measured directly with a moveable quartz crystal microbalance placed near the substrate position. Active oxygen was provided by an Oxford Instruments HD25 radio frequency inductively coupled plasma source equipped with autotuning and alumina discharge tube components as well as electrostatic ion removal plates. Films were deposited on chemically degreased 1 cm² r-plane sapphire substrates with a target growth rate of approximately 1 µm/hr and a thickness range of 600 to 1000 nm. X-ray photoelectron spectroscopy (XPS) was performed using a Phi Versaprobe 5000 equipped with Al Kα source with a hemispherical analyzer. Measurements utilized a pass energy of 23.5eV, corresponding to an instrumental resolution of 0.51 eV, determined from analyzing the Fermi edge and gold foil references. Measurements were performed without any surface preparation.

Here we describe five separate film growth experiments with the goal of obtaining largely SnO as opposed to the more common SnO_2. A summary of the conditions is provided in Table 1, along with the sample identifiers. In all cases, the oxygen plasma source was operated at a forward power of 250 W, which has been shown to provide a mixture of monotomic oxygen as well as energetic molecular oxygen [10].

Table 1. Growth condition summary for the five samples on r-plane sapphire.

Sample Identifier	T_{Sn} (°C)	$T_{substrate}$ (°C)	Oxygen pressure ($\times 10^{-5}$ torr)	Tin deposition rate (Å/sec)	Γ_{Sn} ($\times 10^{14}$ at·cm^{-2}s^{-1})	RF power (watts)
SC17	1170	472	6.	2.2	8.2	250
SC18	1170	423	7.	2.2	8.2	250
SC19	1170	550	6.	2.1	7.8	250
SC20	1050	450	7.	1.5	5.6	250
SC21	1050	450	3.	< 1.5 (variable)	~1.5 (est.)	250

DISCUSSION

It is well-established that nominally stoichiometric growth conditions yield the best quality ZnO thin films when using plasma-assisted molecular beam epitaxy [11]. Since direct measurement of active oxygen is problematic, this growth regime is typically extracted from a series of growth experiments where the Zn flux is increased while holding all other parameters constant. The point at which the growth rate is observed to saturate is typically taken to represent approximately equal Zn and oxygen flux. Below this point, extremely rough films, often characterized by distinct columnar growth, are observed. In the case of SnO_2 growth, similar behavior has been reported, specifically a linear increase in growth rate with increasing Sn flux when growing under oxygen-rich conditions, with an improvement in film quality observed for growth under nominally stoichiometric flux conditions and elevated substrate temperature (~700 °C) [8]. In contrast, however, a reduction in growth rate with further increases in Sn was reported, and attributed to competition through formation of volatile SnO at the growth front [9].

In order to obtain largely SnO containing films, we intentionally grew at lower substrate temperatures, in the range of 423 to 550 °C, and also varied tin flux and oxygen flow rate; RF power was kept constant at 250 W for this set of experiments. Fig. 1a shows the surface appearance of film SC-17 grown at a substrate temperature of 472 °C, an oxygen partial pressure of 6×10^{-5} torr, and a tin flux of approximately 8.2×10^{14} at/cm^2 s. This rough morphology is consistent with growth under oxygen-rich conditions, and is undesirable for device fabrication. Previous reports indicate a preferential growth on r-plane sapphire along the (101) direction, and we see some evidence of this in x-ray diffraction (Fig. 2). However, in the same plot there is clear evidence of other orientations of SnO, along with SnO_2 and a weak feature attributed to metallic tin. Consequently, these conditions are insufficient to yield a purely SnO thin film.

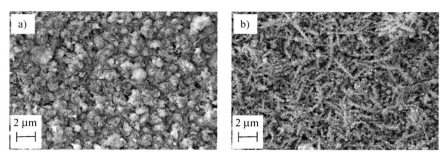

Figure 1 Scanning electron microscopy (SEM) images showing the surface morphology of films (a) SC-17 and (b) SC-19.

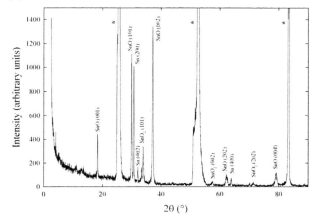

Figure 2 X-ray diffraction of SC-17, showing a largely SnO composition but with poor epitaxial orientation, and clear signs of SnO_2 as well as metallic Sn. Substrate peaks are identified by *.

Further increasing the substrate temperature (to 550 °C) leads to a radically different surface morphology, consisting of what appears to be a random network of nanometer-scale wires (Fig. 1b). X-ray diffraction of this sample (Fig. 3) shows a strong (101) SnO feature, but also many features attributed to various SnO_2 orientations, as well as a weak (001) Sn_3O_4 feature and again evidence of metallic tin. Given the metastable nature of SnO, this is not surprising, although it helps establish an upper substrate temperature limit if high-quality material is desired.

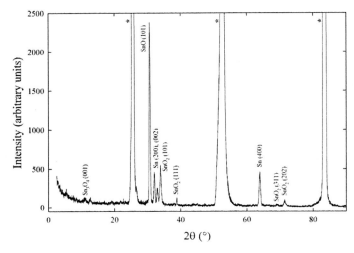

Figure 3 X-ray diffraction of SC-19, showing evidence of epitaxial SnO mixed with SnO_2 and Sn_3O_4, as well as metallic tin.

The last two runs were grown at a lower temperature (450 °C) in an effort to reduce the tendency for SnO_2 to form. SC-20 was grown with approximately the same oxygen conditions as SC-17 and SC-19, but with only about 70% of the Sn flux. Consequently it was also likely grown under oxygen-rich conditions. This film had a much more contiguous and smoother surface morphology, with narrow voids approximately 1 µm long. X-ray diffraction of this sample exhibited only features consistent with SnO formation. The Sn flux was unstable for the final run (SC-21), and as the crucible charge was not depleted, could be due instead to partial oxidation of the surface, a well-documented problem encountered with ZnO PA-MBE growth. Only SnO_2 features were observed for this final film in x-ray diffraction, although it had a smooth surface morphology in SEM.

Analysis of the Sn $3d_{5/2}$ XPS (not shown) indicates that samples SC-17, SC-19 and SC-20 all display mixed Sn^{4+} and Sn^{2+} contributions, consistent with the x-ray diffraction. Of the three, SC-19 has the most Sn^{4+} character, with SC-17 and SC-19 displaying almost comparable $Sn^{2+/4+}$ contributions. The extreme surface sensitivity of the XPS measurements (effective probing depth ~ 2 nm) means that the Sn^{4+} contribution may be enhanced by surface oxidation. The assignments are verified by the corresponding XPS of the valence band region, which shows the emergence of a peak around 2 eV below the Fermi level for SC 17 and 20. This feature has been attributed to Sn 5s - O 2p antibonding states consistent with SnO [12].

Electrical measurements of the sample set proved somewhat challenging, in particular due to the rather extreme sample roughness. Weakly rectifying diodes were achieved on SC-20 and SC-21. SC-20 showed no significant response to ultraviolet excitation, but SC-21 showed a clear response. For the latter sample, a metal-semiconductor-metal (MSM) structure fabricated using gold-capped silver oxide contacts [13] exhibited an unusually high degree of persistant photoconductivity (PPC) upon exposure to ultraviolet light. Additional measurements, including Hall effect, are ongoing, although some evidence for p-type conductivity was observed in one sample.

Detailed photoluminescence (PL) measurements were performed using an argon ion laser (λ = 333 nm, 100 W/cm^2) and Triax 320 spectrometer fitted with a CCD detector, in order to extract additional information concerning the properties of the grown films. PL from SC-20, the one sample which showed the smoothest morphology with majority SnO content, is plotted in Fig. 4 over the temperature range of 4 – 220 K. At 4 K the emission can be deconvoluted into three main features at 2.02, 2.27 and 2.53 eV, respectively. All of these features are far below the expected band gap energy of SnO$_2$, consistent with the x-ray diffraction analysis. All three are significantly lower in energy than that reported by Guo et al. [14], who observed a broad peak centered at 2.82 eV. However, those films were poorly textured films deposited by electron beam evaporation of an SnO$_2$ target, and exhibited p-type conductivity, so direct comparison is difficult.

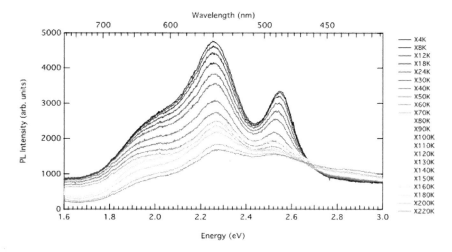

Figure 4 Temperature dependent photoluminescence from SC-20.

CONCLUSIONS

A series of tin oxide films was grown using plasma-assisted molecular beam epitaxy with the goal being to realize high-quality SnO on r-plane sapphire. X-ray diffraction indicates that SnO dominant films are possible through tuning of growth conditions, but SnO_2 formation is difficult to avoid, especially at elevated substrate temperatures. Some evidence of p-type conductivity was observed, as well as persistent photoconductivity. Photoluminescence of one sample was characterized by an absence of high-energy features expected from SnO_2, and instead revealed three broad features in the range of $2 - 2.5$ eV.

ACKNOWLEDGMENTS

This work was supported in part by the State University of New York and the MacDiarmid Institute for Advanced Materials and Nanotechnology. Assistance from J. Aldous, N. Feldberg, B. Keen, I. Tanveer and R. Makin is gratefully acknowledged. L.F.J.P acknowledges support from the Jean-Pierre Mileur Faculty Development Fund.

REFERENCES

1. M. Batzill, U. Diebold, Prog. Surf. Science 79 (2005) 47.
2. T. Nagata, O. Bierwagen, M.E. White, M.Y. Tsai, Y. Yamashita, H. Yoshikawa, N. Ohashi, K. Kobayashi, T. Chikyow, J.S. Speck, Appl. Phys. Lett. 98 (2011) 232107.
3. R.E. Cavicchi, S. Semancik, M.D. Antonik, R.J. Lad, Appl. Phys. Lett. 61 (1992) 1921.
4. R.D. Vispute, V.P. Godbole, S.M. Chaudhari, S.M. Kanetkar, S.B. Ogale, J. Mater. Res. 3 (1988) 1180.
5. R.N. Ghoshtagore, J. Electrochem. Soc. 125 (1978) 110.
6. Y. Ogo, H. Hiramatsu, K. Nomura, H. Yanagi, T. Kamiya, M. Hirano, H. Hosono, Appl. Phys. Lett. 93 (2008) 032113.
7. M. Batzill, J.M. Burst, U. Diebold, Thin Solid Films 484 (2005) 132.
8. M.E. White, M.Y. Tsai, F. Wu, J.S. Speck, J. Vac. Sci. Tech. A 26 (2008) 1300.
9. M.E. White, Molecular Beam Epitaxy and Characterization of Stannic Oxide, Doctoral Thesis, University of California, Santa Barbara (2010).
10. W.C.T. Lee, Doctoral Thesis, University of Canterbury (2008).
11. H.-J. Ko, T. Yao, Y. Chen, S.-K. Hong, J. Appl. Phys. 92 (2002) 4354.
12. N.F. Quackenbush, J.P. Allen, D.O. Scanlon, S. Sallis, J.A. Hewlett, A.S. Nandur, B. Chen, K.E. Smith, C. Weiland, D.A. Fischer, J.C. Woicik, B.E. White, G.W. Watson, L.F.J. Piper, Chem. Mater. 25 (2013) 3114.
13. M.W. Allen, S.M. Durbin, J.B. Metson, Appl. Phys. Lett. 91 (2007) 053512.
14. W. Guo, L. Fu, Y. Zhang, K. Zhang, L.Y. Liang, Z.M. Liu, H.T. Cao, and X.Q. Pan, Appl. Phys. Lett. 96 (2010) 042113.

Epitaxial Growth of (Na,K)NbO$_3$ based materials on SrTiO$_3$ by pulsed laser deposition

T. Hanawa[1,2], N. Kikuchi[1,2], K. Nishio[2], K. Tonooka[1], R. Wang[1], T. Mamiya[1]

[1] National Institute of Advanced Industrial Science and Technology (AIST), 1-1-1 Umezono, Tsukuba, Ibaraki 305-8568, JAPAN
[2] Graduate School of Industrial Science and Technology, Tokyo University of Science, 6-3-1 Niijuku, Katsushika, Tokyo 125-8585, JAPAN

ABSTRACT

Lead-free, piezoelectric (Na,K)NbO$_3$-BaZrO$_3$-(Bi,Li)TiO$_3$ films were epitaxially grown onto (100) SrTiO$_3$ substrate via pulsed laser deposition. The effects of post-annealing temperature on the crystal phases, mosaic spread, and chemical composition of the deposited (Na,K)NbO$_3$ and (Na,K)NbO$_3$-BaZrO$_3$-(Bi,Li)TiO$_3$ films were analyzed. Results indicate the epitaxial growth of (Na,K)NbO$_3$-BaZrO$_3$-(Bi,Li)TiO$_3$ films deposited at an oxygen pressure (P_{O2}) of ≥40 Pa and substrate temperature (T_s) of 800°C. The alkaline-deficiency could be suppressed in the (Na,K)NbO$_3$-BaZrO$_3$-(Bi,Li)TiO$_3$ films deposited at $P_{O2} \geq 70$ Pa. AFM profile of the (Na,K)NbO$_3$ post-annealed at 1000°C indicates the epitaxial growth of film with atomically flat step-terrace structure, while that of the (Na,K)NbO$_3$-BaZrO$_3$-(Bi,Li)TiO$_3$ film post-annealed at 1200°C shows relatively smooth surface with step-terrace structure and several cubic crystals. It was also found that the preferential evaporation of alkaline components could be suppressed by annealing under covered substrate condition.

INTRODUCTION

Piezoelectric materials are being widely used in various electronic devices, such as diesel fuel injectors, printer head of ink jet printers, fishing sonars, and so on. Of the different piezoelectric materials, Pb(Zr,Ti)O$_3$ (PZT) and related lead-containing materials are mainly used in the abovementioned applications because of their high piezoelectric constant (for instance d_{33}) as well as high Curie temperature (T_c) [1,2]. Recently, various regulations such as Restriction of Hazardous Substances/End of Life Vehicles (RoHS and ELV) directives have restricted the use of toxic elements, including lead, in electronic devices and vehicles [3]. This has driven the need for the development of new lead-free piezoelectric materials. To this end, several studies have reported various lead-free piezoelectric materials, including BaTiO$_3$, (Bi,Na)TiO$_3$, and (Na,K)NbO$_3$ with a perovskite structure, and Bi$_4$Ti$_3$O$_{12}$ with a bismuth-layered structure [4-11]. Among them, (Na,K)NbO$_3$ (NKN) and (Na,K)NbO$_3$-based systems have started attracting considerable attention because of their high d_{33} and T_c. In particular, 0.92(Na$_{0.5}$K$_{0.5}$)NbO$_3$-0.06BaZrO$_3$-0.02(Bi$_{0.5}$Li$_{0.5}$)TiO$_3$ exhibits a high T_c of 243°C and d_{33} of 420 pC/N, comparable to those of conventional PZT, making it a potential lead-free alternative piezoelectric material [12]. In our previous study, we reported the epitaxial growth of NKN films onto (100) SrTiO$_3$ substrates and analyzed the intrinsic properties of the NKN films [13]. However, electrical properties including P-E hysteresis curves and piezoelectric constant could not be determined because of the high leakage current between the bottom and top electrodes of the NKN films.
Herein, we report the epitaxial growth of 0.92(Na$_{0.5}$K$_{0.5}$)NbO$_3$-0.06BaZrO$_3$-0.02(Bi$_{0.5}$Li$_{0.5}$)TiO$_3$ (NKN-BZ-BLT) onto (100)STO substrates, and also analyze the effect of

post-annealing conditions on the crystallinity and surface morphology of the obtained NKN and NKN-BZ-BLT films. This allows us to obtain dense structures with high degree of crystallinity.

EXPERIMENTAL

In this study, NKN and NKN-BZ-BLT films were prepared by pulsed laser deposition (PLD) process using $(Na_{0.7}K_{0.3})NbO_3$ and $0.92(Na_{0.5}K_{0.5})NbO_3$-$0.06BaZrO_3$-$0.02(Bi_{0.5}Li_{0.5})TiO_3$ ceramic targets, respectively. The films were deposited onto (100)-oriented single-crystalline $SrTiO_3$ (STO) substrates of thickness 0.5 mm and dimension 10 mm × 10 mm. The substrate STO has a cubic perovskite structure with a lattice constant of a = 0.3905 nm, which is close to those of NKN (a=0.3920 nm) [13] and NKN-BZ-BLT (a=0.4030 nm) [14]. During the PLD process, the deposition chamber was evacuated to ~3 × 10^{-5} Pa using a turbomolecular pump. Ablation was induced using a frequency-quadrupled Nd-YAG laser operating at a wavelength λ = 266 nm. The experimental conditions adopted during the PLD process are as follows: Substrate temperature (T_s) of 800°C, laser repetition rate of 16 Hz, the distance between the target and substrate was maintained as 40 mm, energy density of 12 W/cm^2, and number of laser shots was 172,800. T_s was measured by a thermocouple located near the substrate in a distance of 3 mm. The oxygen pressure (P_{O2}) during the deposition was controlled in the range of 30–90 Pa. The resulting thin films were fully covered by an Yttrium-stabilized-zirconia (YSZ) substrate and air-annealed at 1000–1200°C for 6 h, in order to obtain a dense structure with high crystallinity.

The crystal phase of the films was investigated by X-ray diffraction (XRD, X'Pert Pro, PANalytical) with CuKα radiation, using Bragg-Brentano measurements. In addition, the XRD rocking curves ω-scans were measured to evaluate the mosaic spread of the crystallites in the NKN and NKN-BZ-BLT films. The lattice constant of the films was estimated by measuring the reciprocal space map. AFM images of the films were measured by the scanning probe microscopy (SPI3800, Seiko Instruments) in order to estimate surface morphology. Furthermore, the chemical composition of the films was estimated by using wavelength-dispersive X-ray fluorescence analysis (ZSX, Rigaku).

RESULTS AND DISCUSSION

1. Epitaxial growth of $(Na,K)NbO_3$-$BaZrO_3$-$(Bi,Li)TiO_3$ film

Figure 1 shows the XRD patterns of the NKN-BZ-BLT films prepared at T_s = 800°C under various P_{O2} values. The substrate temperature T_s was chosen to be 800°C, on the basis of the results obtained in our previous study [13]. The NKN-BZ-BLT films prepared at 40 Pa ≤ P_{O2} ≤ 90 Pa showed peaks at 2θ = 22.0° and 2θ = 20.0°, which could be assigned to NKN-BZ-BLT (001) and its Kβ line, respectively. The sharp and intense peaks at 2θ = 22.8° and 20.5° could be assigned to $SrTiO_3$ (001) and its Kβ line, respectively. On the other hand, the film prepared at P_{O2} = 30 Pa showed a peak at 2θ = 29.2°, corresponding to the impurity phase with alkaline-deficiency. In the XRD pattern, the diffraction peaks at 2θ = 23.9° and 24.8° originated due to periodic structure of NKN-BZ-BLT(001) and STO (100). Accordingly, the NKN-BZ-BLT films prepared at 40 Pa ≤ P_{O2} ≤ 90 Pa showed single crystal phase of NKN-BZ-BLT.

Figure 2 shows the full width at half maximum (FWHM) values of the rocking curve for the films deposited under various P_{O2} values, at a substrate temperature T_s of 800°C. The FWHM indicated a minimum value of 0.34° at P_{O2} = 40 Pa. In addition, the FWHM was found to increase with increase in P_{O2}. The FWHM of 0.34°, corresponding to the film prepared at 40 Pa, was relatively larger than that of the as-deposited NKN film (0.23°), as described in the later sections. This result indicates the incorporation of additional ions with different ionic radii, namely A-site ions (Ba^{2+}, Bi^{3+} and Li^+) and B-site ions (Zr^{4+} and Ti^{4+}) onto the Na^+ and K^+ sites, and Nb^{5+} sites of (Na,K)NbO_3 (NKN), respectively, resulting in low crystallinity.

Figure 3 shows the lattice constants of the a- and c-axes for the films as a function of P_{O2} calculated from the reciprocal space map of the XRD. It could be observed that the lattice constant of the a-axis of the films is larger than that of c-axis. The lattice mismatch between the film and $SrTiO_3$ substrate was estimated to be 2.6–3.5%. As evidenced from the figure, both the constants increased with increase in P_{O2}. As the lattice constant of the films is affected by chemical composition of the film, the compositions of the films, as determined by X-ray fluorescence spectroscopy, are presented in Figure 4. In this figure, the composition ratios [Na]/[Nb], [K]/[Nb], and ([Na] + [K])/[Nb] are shown as a function of P_{O2}. Here, we could not detect [Bi], [Li], [Ti], and [Zr], due to the following reasons: 1) The Ti Kα and Zr Kα peaks from the film overlapped with the Ti Kα and Sr Kβ₁ peaks of the $SrTiO_3$ substrate, respectively; 2) Li could not be detected in the XRF system used in this study, as XRF is generally limited to elements with atomic number >16; and 3) [Bi] was too small amount to be detected. As evidenced from the figure, ([Na] + [K])/[Nb] increases with increase in P_{O2}, showing a maximum value at P_{O2} = 70 Pa. However, all the films prepared in this study were found to show alkaline-deficiency, characterized by ([Na] + [K])/[Nb] < 1. The observed increase of both the lattice constant a- and c-axes with increase in P_{O2}, as explained in Figure 3, could be due to the suppression of alkaline-deficiency at high P_{O2}. Although the suppression of alkaline-deficiency was realized at high P_{O2}, the film obtained under P_{O2} = 40 Pa exhibited minimum FWHM. Since the films prepared at high P_{O2} had large lattice constant, the lattice mismatch between the substrate and the films became large, leading to higher FWHM of the rocking curve at high P_{O2}, as evidenced in Figure 2.

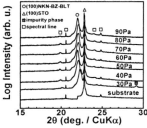

Fig. 1 XRD patterns of NKN-BZ-BLT films obtained under various P_{O2} values (T_s = 800°C)

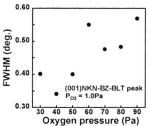

Fig. 2 FWHM of the rocking curve for the NKN-BZ-BLT film

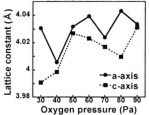

Fig. 3 Lattice constants of the a- and c-axes of the NKN-BZ-BLT films

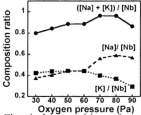

Fig. 4 Chemical composition ratios of the NKN-BZ-BLT thin films obtained under various P_{O2} values (T_s = 800°C)

2. Annealing effect

Figure 5 shows the XRD patterns of the NKN films (a) and the NKN-BZ-BLT films (b), as-deposited and post-annealed at 1000°C, 1100°C, and 1200°C for 6 h. The as-deposited films were prepared at T_s = 800°C, with P_{O2} = 70 Pa for the NKN film and P_{O2} = 40 Pa for the NKN-BZ-BLT film. In case of both the films, no significant difference could be observed among the XRD patterns before and after annealing. However, the difference was determined from the rocking curves of the films.

Figure 6 shows the FWHM of the rocking curve for both the as-deposited and post-annealed NKN and NKN-BZ-BLT films. The FWHM of as-deposited NKN film was estimated to be 0.23°, while that of the as-deposited NKN-BZ-BLT film was found to be 0.34°. The large width (higher FWHM) of the as-deposited NKN-BZ-BLT film could possibly be due to the low crystallinity of the film with a complex composition. It was also found that the FWHM of the NKN-BZ-BLT film decreased with post-annealing treatment, exhibiting the minimum FWHM value corresponding to the annealing temperature of 1100°C. On the hand, the FWHM of the NKN film was found to increase with post-annealing treatment. The observed decrease in the FWHM of NKN-BZ-BLT film with increase in annealing temperature is thought to be due to the enhancement in the crystallinity of the film with annealing treatment. However, the optimum annealing temperature to prepare NKN-BZ-BLT film with high degree of crystallinity may be higher than that of the NKN films.

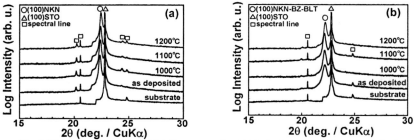

Fig. 5 XRD patterns of the NKN films (a) and NKN-BZ-BLT films (b), as-deposited and post-annealed at 1000°C, 1100°C, and 1200°C. The XRD pattern of SrTiO$_3$ substrate is also included for reference.

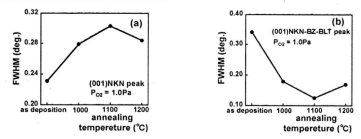

Fig. 6 FWHM of the rocking curve for the NKN films (a) and NKN-BZ-BLT films (b) as-deposited and post-annealed at 1000°C, 1100°C, and 1200°C.

Figure 7 shows the AFM images of the NKN and NKN-BZ-BLT films, as-deposited and post-annealed at 1000°C and 1200°C. The AFM profile of the as-deposited NKN film indicates atomically flat terraces with steps and a few large projections of diameter approximately 500 nm. Similarly, the AFM profile of the NKN film annealed at 1000°C also shows flat terraces with steps, but without the presence of projections. However, with further increase in annealing temperature to 1200°C, the step-terrace structure disappeared completely. On the other hand, the AFM image of the as-deposited NKN-BZ-BLT film shows a rough surface with several projections and holes. Upon annealing at 1000°C, several cubic crystals of diameter approximately 100 nm were found to appear throughout the film surface. The AFM profile of the NKN-BZ-BLT film annealed at 1200°C indicated relatively smooth surface with several cubic crystals. In addition, step-terrace structure could also be observed in the smooth regions. Further increase in the annealing temperature of the NKN-BZ-BLT film resulted in the formation of rough surface. Based on the results obtained by the systematic studies, we could observe the following dependence of the FWHM and the AFM images on the annealing temperature: 1) as-deposited NKN film and that annealed at 1000°C exhibited high crystallinity with a step-terrace structure, although the increase in annealing temperature resulted in low crystallinity; and 2) the crystallinity of the NKN-BZ-BLT film increased with increase in annealing temperature of up to 1100°C, resulting in the formation of a relatively smooth surface with the step-terrace structure.

Finally, the chemical composition of the films before and after annealing was examined by using XRF. Figure 8 shows ([Na] + [K])/[Nb] of the NKN and NKN-BZ-BLT films before and after annealing at 1200°C. In case of both the films, preferential evaporation of alkaline components, resulting in the alkaline-deficient condition, was not found after annealing. This result indicates that the annealing of films under covered substrate condition is effective in controlling the composition of the films.

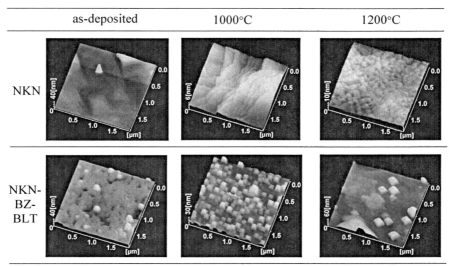

Fig. 7 AFM images of the NKN and NKN-BZ-BLT films, as-deposited and post-annealed at 1000°C and 1200°C.

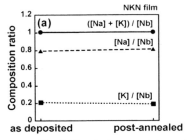
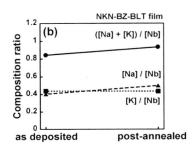

Fig. 8 Chemical composition ratios of the NKN films (a) and the NKN-BZ-BLT films (b) as-deposited and post-annealed at 1200°C

CONCLUSIONS

In summary, $0.92(Na_{0.5}K_{0.5})NbO_3$-$0.06BaZrO_3$-$0.02(Bi_{0.5}Li_{0.5})TiO_3$ films were deposited onto (100)SrTiO$_3$ substrates by pulsed laser deposition technique. We analyzed the dependence of crystal phases, mosaic spread, lattice constant, and chemical composition of the deposited films as a function of the oxygen pressure condition adopted during the film deposition. Results suggest the epitaxial growth of single phase NKN-BZ-BLT films onto (100)STO substrate, deposited under $P_{O2} \geq 40$ Pa and $T_s = 800°C$. With increase in P_{O2}, ([Na]+[K])/[Nb] in the films approached to 1, while lowest mosaic spread was found for the film deposited at $P_{O2} = 40$ Pa. These results were explained by the lattice mismatch between the substrate and the film. Furthermore, the effect of post-annealing temperature on the crystallinity and surface morphology of the NKN and NKN-BZ-BLT films was also examined, in order to arrive at the optimum temperature for obtaining dense structure with high crystallinity. The AFM profile of the NKN film post-annealed at 1000°C showed atomically flat terraces with steps, while that of the NKN-BZ-BLT film post-annealed at 1200°C indicated relatively smooth surface with several cubic crystals. The enhancement in the crystallinity as a result of post-annealing treatment was significant in case of the NKN-BZ-BLT films. It was also found that the annealing of the films under covered substrate condition is effective in controlling the composition of the films.

REFERENCES

1. B. Jaffe, W. R. Cook Jr. and H. Jaffe, *Piezoelectric Ceramics* (Academic Press, London, 1971) p.135-184
2. K. Uchino, *Ferroelectric Devices* (Marcel Dekker, Inc., New York, 2000) p.145-220
3. "Adaptation to scientific and technical progress of Annex II to Directive 200/53/EC (ELV) and of the Annex to Directive 2002/95/EC (RoHS) Final Report", Öko-Institute e.V.-Institute for Applied Ecology, Germany
4. R. Bechman, *J. Acoust. Soc. Am.* **28**, 347 (1956)
5. S. Wada, H. Yasuno, T. Hoshina, S. M. Nan, H. Kanemoto and T. Tsurumi, *J. Appl. Phys.* **98**, 014109 (2005)
6. D. F. K. Henning, C. Metzmacher and S. Schreinemgcher, *J. Am. Ceram. Soc.* **42**, 179 (2001)
7. C. F. Buhrer, *J. Chem. Phys.* **36**, 798 (1962)
8. J. Suchanicz, K. Roleder, A. Kania and J. Handerek, *Ferroelectrics* **77**, 107 (1988)
9. T. Saito, T. Wada, H. Adachi and I. Kanno, *Jpn. J. Appl. Phys.* **43**, 6627 (2004)
10. Y. Nakashima, W. Sakamoto, H. Maiwa, T. Shimura and T. Yogo, *Jpn. J. Appl. Phys.* **46**, L311 (2004)
11. E. C. Subbarao, *J. Am. Ceram. Soc.* **45**, 166 (1962)
12. R. Wang, H. Bando and M. Itoh, IMF-ISAF 2009 Conf. Xi'an, DO-033, (2009)
13. K. Sakurai, T. Hanawa, N. Kikuchi, K. Nishio, K. Tonooka, R. Wang, H. Bando, and H. Takashima, in *Oxide Semiconductors and Thin Films*, edited by A. Schleife, M. Allen, C.B.Arnold, S.M. Durbin, N. Pryds, C.W. Schneider and T. Veal, (Mater. Res. Soc. Symp. Proc 1494, Pittsburgh, PA, 2013) pp.227-232
14. Lattice constant of the NKN- BZ- BLT was estimated from sintered compact for a target.

Structural and electrical properties of LaNiO$_3$ thin films grown on (100) and (001) oriented SrLaAlO$_4$ substrates by chemical solution deposition method

D. S. L. Pontes[1], F. M Pontes[2]*, Marcelo A. Pereira-da-Silva[3,4], O. M. Berengue[5], A. J. Chiquito[5], E. Longo[1,6]

[1]LIEC – Department of Chemistry, Universidade Federal de São Carlos, Via Washington Luiz, Km 235,P.O. Box 676, 13565-905, São Carlos, São Paulo,Brazil
[2]Department of Chemistry, Universidade Estadual Paulista - Unesp, P.O. Box 473, 17033-360, Bauru, São Paulo,Brazil
[3]Institute of Physics of São Carlos, USP, São Carlos, 13560-250, São Paulo, Brazil
[4]UNICEP, São Carlos, 13563-470, São Paulo, Brazil
[5]NanO LaB – Department of Physics, Universidade Federal de São Carlos, Via Washington Luiz, Km 235, P.O. Box 676, 13565-905, São Carlos, São Paulo,Brazil
[6]Institute of Chemistry, Universidade Estadual Paulista – Unesp, Araraquara, São Paulo - Brazil

ABSTRACT

LaNiO$_3$ thin films were deposited on SrLaAlO$_4$ (100) and SrLaAlO$_4$ (001) single crystal substrates by a chemical solution deposition method and heat-treated in oxygen atmosphere at 700°C in tube oven. Structural, morphological, and electrical properties of the LaNiO$_3$ thin films were characterized by X-ray diffraction (XRD), atomic force microscopy (AFM), field emission scanning electron microscopy (FE-SEM), and electrical resistivity as temperature function (Hall measurements). The X-ray diffraction data indicated good crystallinity and a structural preferential orientation. The LaNiO$_3$ thin films have a very flat surface and no droplet was found on their surfaces. Samples of LaNiO$_3$ grown onto (100) and (001) oriented SrLaAlO$_4$ single crystal substrates reveled average grain size by AFM approximately 15-30 and 20-35 nm, respectively. Transport characteristics observed were clearly dependent upon the substrate orientation which exhibited a metal-to-insulator transition. The underlying mechanism is a result of competition between the mobility edge and the Fermi energy through the occupation of electron states which in turn is controlled by the disorder level induced by different growth surfaces.

1 Introduction

Metal-to-insulator transitions (MIT) has been studied in a large variety of systems. These transitions are normally driven by typical parameters such as doping, pressure, temperature and dimensionality (thickness). Metallic perovskites oxides such as La$_{0.5}$Sr$_{0.5}$CoO$_3$ (LSCO) [1], LaNiO$_3$ (LNO) [2,3] and SrRuO$_3$ (SRO) [4] are examples of excellent candidates to explore basic science regarding MIT. These materials can be characterized as strongly electron correlated interactions due to their metallic behavior which in turn is highly dependent upon the dimensionality of the system. Among these materials, the entire class of rare-earth nickelates RNiO$_3$ is known to display MIT which is usually associated with a variation of the unit cell volume as a function of temperature and the radius of the rare-earth material. Of these materials, rare-earth nickelate $LaNiO_3$ is the most interesting, and, surprisingly, $LaNiO_3$ does not present MIT in its bulk form but instead displays a metallic-like behavior in a wide range of temperatures and even in a few Kelvins range [5,6].

$LaNiO_3$ is also the most technologically relevant conductive metallic oxide because it exhibits a simple pseudocubic perovskite structure with a small degree of rhombohedral distortion [7] as well as a relatively simple composition and stoichiometry which enables easier control of the final product during synthesis. Additionally, the pseudocubic perovskite structure has a lattice parameter very close to many monocrystalline substrates, ferroelectric and multiferroic materials. Thus, $LaNiO_3$ also attracted significant attention, especially as a bottom electrode for replacing platinum electrodes in resistance random access memories and ferroelectric or dielectric capacitors when the appreciable metallic conductivity favors integration in electro-electronic devices. In a previous study, we observed that ferroelectric and dielectric properties for highly (100) oriented $Pb_{0.80}Ba_{0.20}TiO_3/LaNiO_3$ heterostructure grown on a $LaAlO_3$ (100) substrate by a chemical solution deposition (CSD) method are considerably better when $LaNiO_3$ is used as a bottom electrode [8]. Recent investigations have shown that single crystal substrates with different conductivity, thickness and orientation also play an important role in chemical and physical properties of $LaNiO_3$ thin films [9].

We report herein the growth, structural, microstructural and electrical properties of nanostructured $LaNiO_3$ thin films grown on (100) and (001) oriented $SrLaAlO_4$ single crystal substrates by using a chemical solution deposition method. The combined results of X-ray diffraction (XRD), atomic force microscopic (AFM), field emission scanning electron microscopy (FE-SEM), and electrical resistivity as temperature function indicated that the growth LNO thin films on different surfaces introduced an interesting and fruitful way to investigate the role of disorder on MIT in this material. The dependence of the observed conduction mechanisms on sample thickness and crystallographic direction is described within a framework where disorder plays a fundamental role. In fact, the properties of $LaNiO_3$ can be artificially tailored to easily control chemical and physical properties.

2 Experimental Procedures

$LaNiO_3$ thin films (therein referred as LNO) were prepared using a chemical solution deposition (CSD) method. The starting chemicals were lanthanum nitrate [$La(NO_3)_3$] and nickel acetate [$Ni(CH_3COOH)_2.4H_2O$] from Alfa Aeasar Co. Water, citric acid and ethylene glycol were used as solvent, chelanting, and polymerizing agents, respectively. The nickel acetate was dissolved in a water solution of citric acid under constant agitation to produce Ni-Citrate. Then, an equimolar amount of lanthanum nitrate was dissolved in distilled water, held at room temperature, and the two solutions were mixed together under constant stirring. After the homogenization of the solution containing La and Ni cations, ethylene glycol was added to promote the citrate polymerization by the polyesterification reaction. With continued heating at 80–90 °C, the solution became more viscous, albeit devoid or any visible phase separation. The viscosity of the deposition solution was adjusted to 15 mPa/s by controlling the water content. Prior to coating, the $SrLaAlO_4$ substrates was cleaned by immersion in a sulfochromic solution followed by rinsing several times in deionized water. Substrates were then dried in an oven at $100°C$ for about one hour.

The polymeric precursor solution was spin-coated on substrates [$SrLaAlO_4$ (100) and $SrLaAlO_4$ (001)] by a spinner operating at a range of 5000-9000 rev./min for 30 s using a commercial spinner (spin-coater KW-4B, Chemat Technology). Two-stage heat treatment was carried out as follows: initial heating at 400°C for 4 h at a heating rate of 5C°/min in an oxygen atmosphere to pyrolyze the organic materials followed by heating at 700°C for 2 h at a heating rate of 5C°/min for crystallization in an oxygen atmosphere.

LNO thin films were then structurally characterized by XRD in the θ-2θ scan mode (steps of 0.02°) which was recorded on a Rigaku D/Max 2400 diffractometer. The thickness of the thin films was characterized using FE-SEM (FEG-VP Zeiss Supra 35, with a secondary electron detector on a freshly fractured film/substrate cross-section. Atomic force microscopy (AFM) was used to obtain a bi-dimensional image reconstruction of the sample surface. These images provided an accurate analysis of the sample surface and the quantification of parameters such as roughness and grain size. A Digital Instruments Multi-Mode Nanoscope IIIa was used in these experiments.

The devices were patterned into Hall bars prepared by standard lithography and chemical etching. The Ohmic contacts were fabricated by depositing 100 nm of Au. A conventional ac four-probe method was used to measure all the electrical parameters. The transport measurements were carried out at different temperatures from 8 to 300 K (±0.1 K) using a closed-cycle helium cryostat and at a pressure lower than 10^{-6} Torr. The resistivity was obtained by using standard low-frequency ac lock-in techniques (f = 13 Hz) with a high noise rejection ratio; dc measurements were also used, but the results remain unchanged. The measurements were taken at both increasing and decreasing temperatures, and no hysteresis was observed in the entire temperature range. Different values for the current used in the experiments were used to avoid nonlinear transport due to high field effects or Joule heating.

3 Results and Discussion

Figure 1 shows X-ray patterns for representative films from three different thicknesses on (100) and (001) oriented $SrLaAlO_4$ substrates. The pseudo-cubic lattice parameter of the bulk LNO is 3.840 Å. $SrLaAlO_4$ substrates have tetragonal structure, which is in this system defined as pseudo-cubic symmetries with lattice parameters a = 3.756 Å. According to bulk LNO pseudo-cubic lattice parameters, the lattice mismatch value is -2.2% compressive strain for (100) and (001) oriented $SrLaAlO_4$ substrates. As we known, strain phenomenon and epitaxial or textured growth are strongly dependent of a critical thickness, deposition method, mismatch between thin film and substrate.

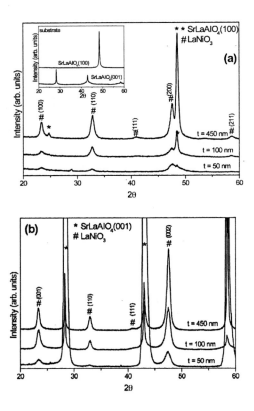

Figure 1. Structural characterization of LaNiO$_3$ thin films of different thicknesses by X-ray diffraction, t = thickness. The inset shows substrates.

However, preferred oriented or textured LNO thin films demonstrated XRD patterns with certain diffraction reflections more pronounced than the others which was strongly dependent on the orientation of the SrLaAlO$_4$ substrate. Moreover, with increasing thickness, intensities of (100), (110) and (200) diffraction peaks, showed distinct behavior on the two different orientations. Significantly, XRD patterns of LNO thin films grown on (001) oriented SrLaALO$_4$ substrate exhibits strong *(00l)* plane family diffractions peaks, suggesting that textured LNO thin films with preferred strong orientation on (001) oriented SrLaALO$_4$ substrate. In addition, bulk powder and polycrystalline thin films samples indicate that the (110) diffraction peak should be the most intense, followed by (200), (111), and (100) diffraction peaks, respectively. This is clearly not the case for the current LNO thin films grown on (001) oriented SrLaALO$_4$ substrate, irrespective of the thickness. Surprisingly, all LNO thin films deposited on (100) oriented SrLaAlO$_4$ substrate (see Figure 1a), exhibited almost same intensity

(110) diffractions peak when compared (100) and (200) diffraction peaks. According to the relative intensity of the (100), (110) and (200) diffraction peaks, a weak texturing along *(h00)* direction has been observed, suggesting that the LNO thin films are almost grown in random orientation and/or lightly textured with preferred weaker orientation. These characteristics on the preferred crystallographic directions can lead to anisotropic properties in LNO thin films (to be discussed in the next paragraphs).

In the present study the orientation factor of the LNO thin films calculated from the X-ray diffraction patterns is shown in Table 1 for *(h00)* and *(00l)* direction as a function of film thickness and substrate orientation.

As clearly shown in Figure 1, with increasing thickness, more and more intense is (110) reflections. This phenomenon should be generated to relieve strain between film/substrate interface, for films with thickness above 50 nm. This characteristic is intrinsic to the chemical solution deposition method where the processing steps involved in forming thick films undergo successive deposition and heat treatment cycles (heating and cooling). In our case, the 50 nm thick films were prepared in a step single deposited directly on (100) or (001) oriented $SrLaAlO_4$ substrates resulting in a strain and preferred orientation effects (see orientation factor values, Table 1). In contrast, to obtain 100 and 450 nm thick films, successive depositions and heat treatment cycles were realized, as consequence more and more dislocations should occur with strain energy released, suggesting preferred orientation degree small compared with 50 nm thick films. A schematic diagram can be draw to illustrates this, as shown in Figure 2.

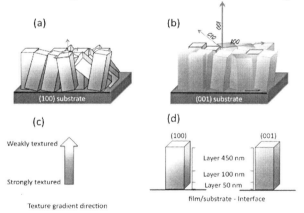

Figure 2. Schematic diagram of the anisotropic texture gradient induced by the substrate's vicinality and film thick. (a) A view of the texture variation on a (100) surface structure, weakly textured. (b) A view of the texture variation on a (001) surface structure, strongly textured. (c) The red and white colors correspond to the strongly and weakly textured states, respectively. (d) Thickness gradient direction

As we known, the preferred orientation mechanism is stronger dependent of the crystallographic planes. Generally, energy minimization considerations and their relation to chemical termination of the substrate surface and film has been exhaustively discussed by many authors in terms of surface energy either from viewpoint theoretical or experimental [10]. In this framework, surface energy is one of the central issues to obtain coherent (epitaxial) or strongly textured growth.

Therefore, the changes in the degree of preferred orientation in our case should be related to the different surface energies characteristic of each substrate surface. Certainly, the surface energy is different when the orientation of the substrate is changed of (001) to (100). Thus, it is possible that more energy (a higher temperature) is required to obtain highly textured LNO thin films on the (100) $SrLaAlO_4$ substrate rather than on the (001) $SrLaAlO_4$ substrate; i.e., the film has a tendency to be highly oriented toward the lowest energy plane. Thus, we propose that the smaller degree of orientation of the LNO film on a (100) oriented substrate may be due to the higher energy of the (100) surface rather than the lattice match between LNO and $SrLaAlO_4$ single crystal substrates. In addition, Figure 1 shows that when film thicknesses are increased from 50 nm to 450 nm, the intensity of the LNO (002) and (200) peaks increases as well, but the full width at half maximum (FWHM) decreases, which indicates that the grain size increased and/or crystallite. In these two cases, we have also estimated the crystallite sizes by using the known Scherrer's equation (see Table 1).

Table 1. Orientation degree (F), average crystallite size (d), lattice parameters (a), and average grain size (t) by AFM of LNO thin films on (100) $SrLaAlO_4$ and (001) $SrLaAlO_4$ single crystal substrate as function thickness.

Orientation	100			001		
Thickness	450nm	100nm	50nm	450nm	100nm	50nm
Parameters						
a (nm)	0.3850	0.3850	0.3849	0.3835	0.3826	0.3821
d (nm)	12.1	9.1	7.5	13.1	9.3	9.1
t (nm)	30	25	15	35	30	20
F	0.50	0.26	0.75	0.84	0.82	0.91

The morphology of LNO thin films was also investigated by AFM. All these samples exhibited a dense and uniform microstructure with spherical-shaped grains. The observed average grain size of the films on (100) and (001) $SrLaAlO_4$ substrates increases from 15 nm to 30 nm and 20 to 35 nm, respectively, suggesting that the grains are formed of at least two or three crystallite, which is qualitatively consistent with XRD results. In addition, films grown on the (100) surface have a smaller average grain size than films grown on the (001) surface. All the thin films have smooth surfaces with root-mean-squared (RMS) roughness below 5 nm.

As a matter of fact, the orientation of the substrate directly influences the growth rate of the grains and the degree of orientation. Along with the film thicknesses, we believe that these parameters play a key role in electrical properties of LNO thin films which will be explored in this work.

Cross section analysis of these films were performed by FE-SEM (FEG-VP Zeiss Supra 35) and revealed that samples display similar thickness of approximately 50, 100 and 450 nm on both (100) and (001) $SrLaAlO_4$ single crystal substrates.

The temperature dependent resistivity data obtained for samples of different thicknesses grown on substrates with (100) and (001) orientations are shown in Figure 3. Both thickness and substrate orientation induce dramatic changes in the electron transport in these samples. Beginning with the resistivity dependence on the thickness of the films, samples grown on the (100) surface exhibit a monotonic activated-like behavior for all film thicknesses such as the behavior observed in insulating or semiconducting materials (see Figure 3). In addition to the crystalline quality of the samples, the presence of disorder cannot be avoided; as a consequence, the electron potential is also disordered.

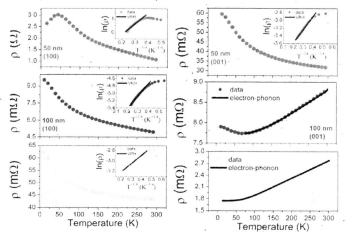

Figure 3. Temperature dependence of the electrical resistivity of the LaNiO$_3$ samples grown on (100) SrLaAlO$_4$ and (001) single crystal substrate and with different thicknesses: 50 nm, 100 nm and 450 nm.

Any carrier subjected to a random potential is unable to move freely through the system if either potential fluctuations exceed a critical value or the electron energy is lower than the characteristic value of the potential fluctuations, leading to Anderson-type localization and hence explain the observed insulating behavior. Considering the motion of charges in disordered potentials, the expected conduction mechanism (metallic) will change its character to an activated process such as the variable range hopping (VRH) reported by Mott, for instance. In this model, phonons are required to conserve energy during a hop from site to site; a higher phonon density at a higher temperature increases the hopping rate and thereby decreases the resistivity. The VRH mechanism is described by the equation

$$\rho(T) = \rho_0 \exp\left(\frac{T_0}{T}\right)^{1/4}, \qquad (1)$$

where $T_0 = \dfrac{5.7\alpha^3}{k_B N(E_F)}$, N(E$_F$) is the density of states at the Fermi level and α^{-1} is the localization length. The VRH mechanism normally occurs only in the temperature region where the energy is insufficient to excite the charge carrier across the Coulomb gap between two states. Hence conduction takes place by the hopping of a small region (k$_B$T) in the vicinity of the Fermi level where the density of states remains almost a constant (m=1/4). This condition is fulfilled when the temperature is sufficiently small or when the energy states are uniformly distributed.

The fitting of eq. (1) to the experimental data is shown in Figure 3 (see insets). The agreement of both theoretical and experimental curves for all samples in a large range of temperatures (40 K < T < 260 K) confirms that the transport in that region is governed by the variable range hopping process.

Surprisingly, samples grown on (001) SrLaAlO$_4$ single crystal substrates show a distinct behavior, Figure 3. Changes in the electrical behavior of the films, as a function of temperature as the thickness is varied is clearly evident in Figure 3. The 50 nm thick sample still shows a localized behavior for transport which agrees with the VRH model (see inset) for $\alpha^{-1} = 10$ Å. However, for LNO thin film with a thickness of 50 nm and growth on (001) SrLaAlO$_4$ single crystal substrate the behavior is insulating over the entire temperature range. Boris et al [2] reported a similar result to superlattices composed of thick consecutive layers of LaNiO$_3$ and LaAlO$_3$ on (001) SrLaAlO$_4$ single crystal substrate growth by pulsed laser deposition. The breakpoint between conduction regimes is at 100 K and is evident in the 100 nm thick samples when the insulating behavior is replaced by a metallic behavior as expected in MIT. Above this limiting value, the conduction mechanism shows a metallic character; as the temperature and phonon excitation increase, the amount of scattering events experienced by the conduction electrons increase as well which results in greater (metal) resistance. Below 100 K, the resistivity data show a localized character which in turn has the very same origin as observed in samples grown on (100) substrates.

The Ioffe-Regel criterion for MIT takes place when $k_F \ell = 1$ where k_F is the Fermi wavelength and ℓ is the mean free path. This requirement is fulfilled by samples grown on (001) substrates: $k_F \ell \sim 0.4$ (100 nm thick) and $k_F \ell \sim 0.9$ (450 nm thick). The increase in the resistance for temperatures higher than 100 K can be understood by considering the electron-phonon scattering mechanism (at low disorder limits). This behavior agrees very well with the Bloch-Grüneisen theory for the dependence of the electrical resistance on the electron-acoustic phonon scattering mechanism. Thus, the resistance is described by

$$\rho(T) = \rho_0 + A\left(\frac{T}{\Theta_D}\right)^n \int_0^{\frac{\Theta_D}{T}} \frac{z^n e^z}{\left(e^z - 1\right)^2} dz, \qquad (2)$$

where A and ρ_0 are constants and n usually ranges from 3 to 5 when the electron-phonon interaction is mainly responsible for scattering events; Θ_D is the Debye temperature. When the $T \geq \Theta_D$, all phonon modes are excited, and the resistivity increases linearly with increasing temperature. However, when $T \leq \Theta_D$, phonon scattering is mostly dominated by one-phonon normal processes where only small scattering vectors contribute. The fitting of experimental data using eq. (2) revealed n = 5 and Θ_D = 733 K (100 nm sample) and 481 K (450 nm sample). As the temperature and phonon excitation increase, the amount of scattering events experienced by conduction electrons increase as well which results in greater resistance. From the literature, $\Theta_D = 385$ K ~ 500 K. The observed Debye temperature is a definitive proof at electron-phonon interaction and then a metallic phase are dominating the conduction mechanism in those samples. Importantly, the 100 nm thick sample has a high value of Θ_D due to the clear presence of two different coexisting conduction mechanisms. To confirm the metallic phase, a thicker sample (1000 nm) was fabricated; as expected, Θ_D = 478 K (not shown here).

4 Conclusions

In this work, crystalline LNO thin films were obtained by chemical solution deposition method on both (100)- and (001)-oriented SrLaAlO$_4$ single crystal substrates.

It was shown that the microstructure, structural and electric properties measurements is greatly dependent substrates crystallographic orientation and thin films thickness. The X-ray diffraction analysis indicated high degree preferential orientation along the (001) direction for LNO thin films on (001)-oriented $SrLaAlO_4$ substrate than on (100)-oriented $SrLaAlO_4$ substrates. The AFM images reveal that the LNO thin films growth onto both (100) and (001) $SrLaAlO_4$ single crystalline substrates exhibited homogenous grain distribuion, atomically smooth surface and pinhole-free. A general picture of the transport properties was constructed. In fact, the underlying mechanism discussed is based on the interplay of disorder, crystallographic orientation and thickness. We argue that the degree of orientation for different samples implies a high level of disorder; electron localization then occurs. MIT is a result of the competition between a mobility edge and the Fermi energy through the occupation of electron states which in turn is controlled by the disorder level induced by different growth surfaces. The present results further suggest that by choosing an appropriate thickness, substrate and crystallographic orientation, we can still achieve a $LaNiO_3$ thin film with metallic or insulating behavior.

ACKNOWLEDGMENTS

This work was financially supported by the Brazilian agencies FAPESP, CNPq and CAPES. We thank CEPID/CMDMC/INCTMN. FAPESP process n°. 08/57150-6 and n°.11/20536-7.

References

1. R. J. Zeches, M. D. Rossell, J. X. Zhang, A. J. Hatt, Q. He, C.-H. Yang, A. Kumar, C. H. Wang, A. Melville, C. Adamo, G. Sheng, Y.-H. Chu, J. F. Ihlefeld, R. Erni, C. Ederer, V. Gopalan, L. Q. Chen, D. G. Schlom, N. A. Spaldin, L. W. Martin, R. Ramesh, Science **326**, 977-980 (2009).
2. A. V. Boris, Y. Matiks, E. Benckiser, A. Frano, P. Popovich, V. Hinkov, P. Wochner, M. Castro-Colin, E. Detemple, V. K. Malik, C. Bernhard, T. Prokscha, A. Suter, Z. Salman, E. Morenzoni, G. Cristiani, H.-U. Habermeier, B. Keimer, , Science **332**, 937-940 (2011).
3. H. German, S. Nicola, Science **332**, 922-923 (2011).
4. J. M. Rondinelli, N. M. Caffrey, S. Sanvito, N. A. Spaldin, Phys Rev B **78**, 155107-155122 (2008).
5. N. Gayathri, A. K. Raychaudhuri, X. Q. Xu, J. L. Peng, R. L. Greene, J. Phys.: Cond. Matt. **11**, 2901(1999)
6. P. Lacorre, J. B. Torrance, J. Pannetier, A. I. Nazzal, P. W. Wang, T. C. Huang, J. Solid State Chem **91**, 225-237 (1991).
7. K. Ueno, T. Yamaguchi, W. Sakamoto, T. Yogo, K. Kikuta, S-i. Hirano, Thin Solid Films **491**, 78-81 (2005).
8. F. M. Pontes, E. R. Leite, G. P. Mambrini, M. T. Escote, E. Longo E, J. A. Varela, Appl Phys Lett **84**, 248-250 (2004).
9. R. Scherwitzl, S. Gariglio, M. Gabay, P. Zubko, M. Gibert, J. M. Triscone, Metal-Insulator Transition in Ultrathin $LaNiO_3$ Films, Phys Rev Lett **106**, 246403-246407 (2011).
10. X. Wang, S. Olafsson, P. SandstroÈm, U. Helmersson, Thin Solid Films **360**, 181-186 (2000).

Optical and Electrical Characterization

Native point defects in multicomponent transparent conducting oxides

Altynbek Murat and Julia E. Medvedeva
Department of Physics, Missouri University of Science & Technology, Rolla, MO 65409, USA

ABSTRACT

The formation of native point defects in layered multicomponent InAMO$_4$ oxides with A^{3+}=Al or Ga, and M^{2+}=Ca, Mg, or Zn, is investigated using first-principles density functional calculations. We calculated the formation energy of acceptor (cation vacancies, acceptor antisites) and donor (oxygen vacancy, donor antisites) defects within the structurally and chemically distinct layers of InAMO$_4$ oxides. We find that the antisite donor defect, in particular, the A atom substituted on the M atom site (A_M) in InAMO$_4$ oxides, have lower formation energies, hence, higher concentrations, as compared to those of the oxygen vacancy which is know to be the major donor defect in binary constituent oxides. The major acceptor (electron "killer") defects are cation vacancies except for InAlCaO$_4$ where the antisite Ca$_{Al}$ is the most abundant acceptor defect. The results of the defect formation analysis help explain the changes in the observed carrier concentrations as a function of chemical composition in InAMO$_4$, and also why the InAlZnO$_4$ samples are unstable under a wide range of growing conditions.

INTRODUCTION

Multicomponent transparent conducting oxides [1-5] with the composition of both post-transition metals (In, Ga, and Zn) and main-group light-metals (Al, Mg, and Ca) are highly attractive since the presence of the main-group light-metal cations (i) may stabilize the multi-cation structure; (ii) allows for a broader optical transmission window due to a larger band gap, (iii) changes the work function via tunable band edges; and (iv) also helps control the carrier content while preserving the carrier mobility [3].

In order to determine the carrier generation mechanism(s) and the role of each constituent cation in the carrier generation and transport, all possible native point defects in a multicomponent oxide must be carefully studied. In this work, first-principles density functional approach is employed to systematically calculate the formation energies of possible acceptor and donor point defects in InGaZnO$_4$ as well as in light-metal containing InAlZnO$_4$, InGaMgO$_4$, and InAlCaO$_4$. Next, we determine the equilibrium defect and electron densities as a function of growth temperature and oxygen partial pressure. The results reveal that the oxygen vacancies, long believed to be the carrier source in these oxides, are scarce, and cation antisite defects are the major electron donors in the conductive oxides. The proposed carrier generation mechanism helps explain the observed behavior of the conductivity and carrier concentration as a function of composition. The defect analysis also reveals the reasons for instability of InAlZnO$_4$ samples.

THEORY

Ab-initio full-potential linearized augmented plane wave method (FLAPW) [6,7] with the local density approximation (LDA) and with the screened-exchanged LDA [8] is employed for the investigation of the defect formation energies and the electronic properties of InAMO$_4$ oxides. Cutoffs for the basis functions, 16.0 Ry, and the potential representation, 81.0 Ry, and expansion in terms of spherical harmonics with $l < 8$ inside the muffin-tin spheres were used. The muffin-tin radii of multicomponent oxides are as follows: 2.3 to 2.6 a.u. for In and Ca; 1.7 to

2.1 a.u. for Ga, Zn, and Al; and 1.45 to 1.8 a.u. for O atoms. Summations over the Brillouin zone were carried out using at least 23 special **k** points in the irreducible wedge.

The investigated InAMO_4 oxides have rhombohedral R-$3m$ layered crystal structure of YbFe$_2$O$_4$ type [9-11], Fig. 1(a). In these compounds, In^{3+} ions have octahedral coordination with the oxygen atoms and reside in 3(a) position (Yb), whereas both A^{3+} (Al or Ga) and M^{2+} (Ca, Mg, or Zn) ions reside in 6(c) position (Fe), Fig. 1, and are distributed randomly. Because of the different ionic radii and the valence state of the cations in the $AMO_{2.5}$ double layer, the A and M atoms have different z component of the internal site position 6(c). The optimized structural parameters for every structure under consideration can be found in our previous work [12].

To model isolated point defects in the InAMO_4 compounds, a 49-atom supercell was used with the lattice vectors (30-2), (-112), and (02-1), given in the units of the rhombohedral primitive cell vectors [13]. These supercells result in similar defect concentrations, namely, 1.6--1.8x10^{21} cm^{-3}, and, hence, similar distances between the defects ~10 Angstrom. As mentioned above, the layered crystal structure of InAMO_4 oxides has two chemically and structurally distinct layers, $AMO_{2.5}$ and InO$_{1.5}$, which alternate along the [0001] direction. Depending on the layer and the different nearest-neighbor coordination, there are several structurally non-equivalent sites for point defects. As an example, Figure 1(b) shows four possible antisite defect locations (A_M, In$_M$, M_A, and M_{In}), considered for the InAMO_4 oxides.

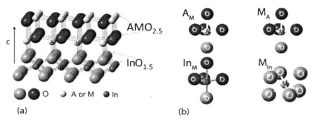

Figure 1. Crystal structure of InAMO_4. (a) One of the three similar blocks that construct the conventional unit cell when stacked along the z direction, is shown. (b) Four possible structurally non-equivalent antisite defects with different nearest-neighboring atoms in the layered multicomponent InAMO_4 oxides.

The formation energy of a defect in a particular charge state (modeled using a corresponding background charge) is calculated as a function of the Fermi level and the corresponding chemical potential [14-16]. The chemical potential was taken with respect to the chemical potential of the elementary metals or the O$_2$ molecule, whereas a deviation from the elemental chemical potential was determined by the specific growth conditions that depend on the temperature and oxygen partial pressure [14] as well as the relevant thermodynamic stability conditions. Specifically, the elemental chemical potentials should have the values that (i) prevent the precipitation of the elements In, A, M, and O; (ii) prevent the formation of the binary compounds, such as In$_2$O$_3$, A_2O$_3$, or MO; and (iii) require a stable InAMO_4 host. The heat of formation for the four representative oxides, InGaZnO$_4$, InAlZnO$_4$, InGaMgO$_4$, and InAlCaO$_4$, calculated with respect to the bulk orthorhombic Ga, tetragonal In, hexagonal Zn or Mg, and cubic Al or Ca, gives --11.28 eV, --14.60 eV, --14.50 eV, and --15.40 eV, respectively. Combined with the heat of formation of binary constituents, the above values suggest that at zero temperature, the formation of multicomponent oxides is impossible without the formation of the corresponding binary phases. Since the multicomponent oxides are stable above 1000 K [9-11], the entropy term must be taken into consideration [16,17]. As a result, the available range for the elemental chemical potentials in the case of quaternary InAMO_4 materials is a very narrow three-dimensional volume determined by the stability conditions of three binary oxides, projected onto the corresponding InAMO_4 surface plane. Our obtained chemical potential values for the four representative compounds in oxygen-rich (pO_2=1 *atm*) and oxygen-poor (pO_2=0.0001 *atm*) conditions are shown in Table 1.

Table 1. Calculated chemical potential values, $\Delta\mu$ in eV, for O-rich (pO_2=1 *atm*) and metal-rich (pO_2=0.0001 *atm*) conditions at the growing temperature T=1000 K for InAMO_4 oxides.

T_g=1000 K	Oxygen-rich (pO_2=1 *atm*)				Oxygen-poor (pO_2=0.0001 *atm*)			
InAMO_4	$\Delta\mu_{In}$	$\Delta\mu_A$	$\Delta\mu_M$	$\Delta\mu_O$	$\Delta\mu_{In}$	$\Delta\mu_A$	$\Delta\mu_M$	$\Delta\mu_O$
InGaZnO$_4$	-2.4	-2.5	-2.3	-1.1	-1.8	-1.9	-1.9	-1.5
InGaMgO$_4$	-2.4	-2.5	-4.6	-1.1	-1.9	-1.9	-4.2	-1.5
InAlZnO$_4$	-2.2	-5.1	-2.2	-1.2	-1.5	-4.4	-1.7	-1.7
InAlCaO$_4$	-2.0	-4.9	-4.6	-1.4	-1.3	-4.2	-4.1	-1.9

DISCUSSION

The formation of native point defects is investigated in the three representative multicomponent InAMO_4 compounds with none, one, and two light-metal constituents. Below, we discuss the role of chemical composition, atomic coordination, and metal-oxygen bond strengths in the formation and stability of native point defects in InAMO_4. In addition to the cation and anion vacancies, there are several possible antisite defects in within the layered structure of InAMO_4 with In^{3+}, A^{3+}, and M^{2+}: those include n-type (donor), p-type (acceptor), and charge-neutral defects. We consider possible charge-producing antisites: In$_M$, A_M (donor) and M_{In}, M_A (acceptor), Fig. 1. Our calculated formation energies for the considered antisite defects in InAMO_4 oxides are plotted in Fig. 2 as a function of the Fermi level.

Donor defects. As found and described in details in Ref. [18], the calculated formation energy of an oxygen vacancy varies with the defect site location in InAMO4. The formation energy of an oxygen vacancy is primarily determined by the strength of the metal-oxygen bonds (i.e., the corresponding binary oxide heat of formation). At the same time, it was shown that the defect formation energy is also affected by the ability of the neighbor cation(s) to form stable low-coordinated structures and the ability of the lattice to relax when adjusting to a new environment created by the defect. In Figure 2, only the configurations with the lowest defect formation energies (that correspond to an oxygen vacancy in the InO$_{1.5}$ layer [18]) are shown. It can be seen from Figure 2 that independent of the chemical composition of the oxides, the oxygen vacancies exhibit similar formation energies and, hence, similar defect concentrations could be expected for these oxides. However, it has been shown earlier that oxygen vacancy in InGaZnO$_4$ is a deep defect and cannot explain the observed persistent conductivity in this material [19-20]. Moreover, oxygen defect alone cannot explain the observed variation in the carrier concentration in InAMO_4 [21], as well as the observed conductivity behavior in (In,Ga)O$_3$(ZnO)$_k$ systems grown under In- vs Ga-rich conditions.

Now, we investigate the formation of antisite defects (cation disorder) in InAMO_4 compounds. We find that in all oxides considered, donor antisites A_M and In$_M$ have lower formation energies than the oxygen vacancy, even under extreme oxygen-poor conditions. Hence, the results reveal that oxygen vacancies, long assumed to be the carrier source in main-group metal oxides, are scarce, and that cation antisite defects are the major electron donors in these multi-cation oxides.

Figure 2 shows that the most abundant donor defect is the A_M antisite defect, whereas the In_M defect has higher formation energy. This trend in the formation energies of the antisite defects correlate well with the experimental heat of formation of the corresponding constituent binary oxides. The experimental values of the heat of formation of binary oxides per oxygen increase in the following order: In_2O_3 (-3.21 eV) > ZnO (-3.60 eV) > Ga_2O_3 (-3.73 eV) > Al_2O_3 (-5.78 eV) > MgO (-6.20 eV) > CaO (-6.57 eV). This means that In-O bonds correspond to the set of the metal-oxygen bonds that would be easiest to break whereas breaking the Ca-O bonds would require more energy. At the same time, one can expect an energy preference for creating stronger bonds during the antisite defect formation.

More specifically, in $InGaMO_4$ with M=Zn or Mg, the donor antisite Ga_M has lower formation energy compared to In_M. Indeed, while breaking the same number of M-O bonds in both cases, the creation of a stronger Ga-O bond is preferred over the weaker In-O bond. Similarly, in $InAlMO_4$ with M=Zn or Ca, the antisites Al_M are more likely to form as compared to In_M defects due to the energy gain associated with the formation of stronger Al-O bonds.

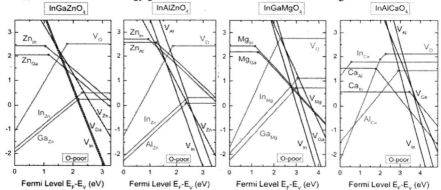

Figure 2. Calculated formation energies of native point defects in $InAMO_4$ for O-poor conditions, i.e., pO2=0.0001 atm and at the growing temperatures of 1023 K – 1223 K. The band gaps calculated within non-local density approximation [8,11] are used, namely, 3.29 eV, 3.48 eV, 4.31 eV, and 4.87 eV, for $InGaZnO_4$, $InAlZnO_4$, $InGaMgO_4$, and $InAlCaO_4$, respectively. The solid squares represent the transition points in each defect.

Acceptor defects. Similar to the donor defects, the formation energy of the antisite acceptor defects is lower than that of the *neutral* cation vacancies in $InAMO_4$ compounds. Figure 2 shows that the acceptor M_A antisite defects are present at higher concentrations than the acceptor M_{In} defects in $InGaZnO_4$, $InGaMgO_4$, and $InAlZnO_4$. This finding cannot be explained using the oxide heat of formation. One of the possible reasons for the higher formation energy of the Zn_{In} defect as compared to the Zn_M antisite in $InGaZnO_4$ and $InAlZnO_4$ is the cation preference for specific oxygen coordination: the structure with six-coordinated Zn (rocksalt

ZnO) is unstable, and Zn has a preference for four-coordinated environment (wurtzite ZnO). For Ca this trend is reversed: in the ground state phase, CaO is octahedrally coordinated (rocksalt CaO), hence, Ca_{In} is more likely to occur than Ca_{Al}. Indeed, the Ca_{In} antisite has lower formation energy than the Ca_{Al} defect InAlCaO$_4$, Figure 2.

In addition to the cation preference to form stable structures with specific oxygen coordination, lattice relaxation caused by the defect may also affect its formation energy. We find that the structural relaxation increases with the number of light-metal constituents. For example, the average distance change for M_A antisite defect in InAlCaO$_4$ is twice larger than that in InAlZnO$_4$ and almost three times larger than that in InGaZnO$_4$. Moreover, for all oxides considered, the relaxation around M_A defect is significantly larger (4-20%) than that around M_{In} antisite (0-6%). This may explain the energy preference of the M_A defect over the M_{In} one.

In order to determine charge compensation mechanisms and the equilibrium defect and carrier concentrations, possible charged states of the acceptor defects must be considered. Our calculated formation energies for the cation vacancies in InAMO_4 correlate with the experimental heat of formation of the constituent binary oxides, i.e., with the metal-oxygen bond strengths. Indeed, the formation of V_{In} is more likely than that of V_{Al}, and the formation of V_{Zn} is lower than that of V_{Ca}. Similar to the antisite defects, the respective formation energies are significantly affected by the large structural relaxation around the vacancy defects. The latter appears due to the presence of several cations of different ionic size, valence, metal-oxygen bond strength, and oxygen coordination near each defect. Indeed, a larger atomic relaxation around V_{In} is found in InGaZnO$_4$ (10—17 %) as compared to that in In2O3 (5—12 %). As a result, the cation vacancies in InAMO_4 result in much lower formation energies than the formation of cation vacancies in the constituent binary oxides [15].

Figure 2 shows that close to the conduction band bottom, the formation energies of the cation vacancies become lower than those for the acceptor antisites. Importantly, the formation of cation vacancies is sensitive to the oxygen-poor/rich conditions; hence, the charge compensation can be controlled via the oxygen partial pressure during the oxide growth. The presence of cation vacancies pushes the equilibrium Fermi level away from the conduction band bottom and deeper into the band gap. We find E_F to be located 1.2 eV, 1.4 eV, 1.6 eV, or 2.1 eV below the conduction band bottom in InGaZnO$_4$, InAlZnO$_4$, InGaMgO$_4$, or InAlCaO$_4$, respectively. This illustrates that the larger the content of the light-metal constituents, the harder it is to achieve a shallow donor.

CONCLUSIONS

Using the obtained formation energy of possible acceptor and donor defects, the equilibrium Fermi energy can be calculated based on the charge neutrality condition for each oxide at a particular growth temperature and oxygen partial pressure. The resulting equilibrium defect and carrier concentrations have been reported for InGaZnO$_4$ [16]. In particular, it was found that the electron concentration follows $Log(pO_2)^{-1/4}$ dependence which is in excellent agreement with the observed dependence of the conductivity in InGaZnO$_4$ [16]. Moreover, the proposed carrier generation mechanism in InGaZnO$_4$ explains the observed intriguing behavior of the conductivity in In-rich vs Ga-rich oxides of the $(In,Ga)O_3(ZnO)_k$ systems with k>1.

Similar to InGaZnO$_4$, the A_M donor antisites have the lowest formation energy in InGaMgO$_4$, InAlZnO$_4$, and InAlCaO$_4$. However, the equilibrium defect analysis in these oxides reveals the following differences in the carrier generation in these oxides:

(i) The low formation energy of the acceptor antisites, Mg_{Ga} and Mg_{In}, leads to an additional charge compensation of the donor defects in $InGaMgO_4$. As a result, the calculated equilibrium electron concentration is reduced by almost two orders of magnitude when Zn is replaced with Mg in $InGaMO_4$. This finding is in excellent agreement with the observed carrier density of 5×10^{19} cm^{-3} in $InGaZnO_4$ and 1×10^{18} cm^{-3} in $InGaMgO_4$ [21].

(ii) In $InAlCaO_4$, the formation energy of the acceptor antisite defects is lower than that of the donor antisites. Moreover, because the concentration of the major donor defect (Al_{Ca}) is similar to that of the major acceptor defect (Ca_{Al}) and both defects occur within the same structural layer, no free carriers are created in $InAlCaO_4$ – even under the extreme oxygen-poor conditions, i.e., at $pO_2=1\times10^{-11}$ atm.

(iii) In $InAlZnO_4$, within the whole range of the possible oxygen partial pressure values ($pO_2=1\times10^{10}$ atm to 1×10^{-14} atm), the equilibrium concentration of Al_{Zn} antisite defect exceeds the maximum possible defect concentration, $n_{max}=1.2\times10^{22}$ cm^{-3}, which is determined as the number of possible sites for this defect in the supercell. Hence, the material will not be stable. The results explain why the $InAlZnO_4$ samples were unstable under a wide range of growing conditions due to the formation of Al-rich Al_2ZnO_4 secondary phases [21].

ACKNOWLEDGMENTS

The work was supported by the National Science Foundation (NSF) grant DMR-1121262. Computational resources are provided by the NSF-supported XSEDE program.

REFERENCES

[1] D. S. Ginley and C. Bright, MRS Bulletin. 25, 15 (2000).
[2] A. Facchetti and T. Marks, Transparent Electronics: From Synthesis to Applications (John Wiley & Sons, New York, 2010).
[3] D.S. Ginley, H. Hosono, D.C. Paine, Handbook of Transparent Conductors (Springer, 2011).
[4] J. E. Medvedeva, Appl. Phys. A 89, 43 (2007).
[5] A. Walsh, J. D. Silva, and S. Wei, J. Phys.: Condens. Matter 23, 334210 (2011).
[6] E. Wimmer, H. Krakauer, M. Weinert, and A. J. Freeman, Phys. Rev. B 24, 864 (1981).
[7] M. Weinert, E. Wimmer, and A. J. Freeman, Phys. Rev. B 26, 4571 (1982).
[8] R. Asahi and W. Mannstadt and A. J. Freeman, Phys. Rev. B 59, 7486 (1999).
[9] V. K. Kato, I. Kawada, N. Kimizuka, and T. Katsura, Z. Krist 141, 314 (1975).
[10] N. Kimizuka and T. Mohri, J. Solid State Chem 60, 382 (1985).
[11] T. M. N. Kimizuka and Y. Matsui, J. Solid State Chem. 74, 98 (1988).
[12] A. Murat and J. E. Medvedeva, Phys. Rev. B 85, 155101 (2012).
[13] J. E. Medvedeva, Europhys. Lett. 78, 57004 (2007).
[14] J. Osorio-Guillen, S. Lany, S. V. Barabash, A. Zunger, Phys. Rev. Lett. 96, 107203 (2006).
[15] S. Lany and A. Zunger, Phys. Rev. Lett. 98, 045501 (2007).
[16] A. Murat, A. Adler, T.O. Mason, J. E. Medvedeva, J. Amer. Chem. Soc. 135, 5685 (2013).
[17] H. Peng, J.-H. Song, E.M. Hopper, Q. Zhu, T.O. Mason, A.J. Freeman, Chem. Mat. 24, 106 (2012).
[18] A. Murat and J. E. Medvedeva, Phys. Rev. B 86, 085123 (2012).
[19] H. Omura, H. Kumomi, K. Nomura, T. Kamiya, M. Hirano, H. Hosono, J. Appl. Phys. 105 (2009).
[20] J. E. Medvedeva and C. L. Hettiarachchi, Physical Review B 81, 125116 (2010).
[21] M. Orita, M. Takeuchi, H. Sakai, and H. Tanji, Jpn. J. Appl. Phys. 34, L1550 (1995).

Electronic Transport Characterization of BiVO$_4$ Using AC Field Hall Technique

Jeffery Lindemuth[1], Alexander J. E. Rettie[2], Luke G. Marshall[4], Jianshi Zhou[4], and C. Buddie Mullins[2,3,4]

1 Lake Shore Cryotronics, Westerville Ohio 43082 (USA)
2 McKetta Department of Chemical Engineering, The University of Texas at Austin, TX 78712 (USA)
3 Center for Electrochemistry, Department of Chemistry and Biochemistry, The University of Texas at Austin, TX 78712 (USA)
4 Materials Science and Engineering Program, Texas Materials Institute, Department of Mechanical Engineering, The University of Texas at Austin, TX 78712 (USA)

ABSTRACT

Bismuth vanadate (BiVO$_4$) is a photoelectrode for the oxidation of water. It is of fundamental importance to understand the electrical and photoelectrochemical properties of this material. In metal oxides, the electronic transport is described by the small polaron model, first described by Mott. In this model, the resistivity varies with temperature as $\rho(T) \propto T e^{(E_a/(k_B T))}$, where E_a is the hopping activation energy, k_B is the Boltzmann constant and T is the absolute temperature. Resistivity measurements confirm that small polaron hopping dominates in temperature ranges from 250 K to 300 K. In addition measurements from 175K to 250K show the variable range hopping dominates the transport. To this end, the electronic transport properties of BiVO$_4$ single crystal were characterized using resistivity measurements and Hall effect measurements over temperatures ranging from 175 K to 300 K.

INTRODUCTION

A promising method for storing solar energy is to directly produce hydrogen gas by using photoelectrochemical (PEC) cell. The PEC cell will directly split water into hydrogen and oxygen. [1,2] The choice of the photoelectrode is critical to the performance of the PEC cell. An ideal photoelectrode is a stable material composed of abundant elements. The photoelectrode should have good light absorption and high quality charge transport properties. Metal oxides, such as titania (TiO$_2$) [3], hematite (α-Fe$_2$O$_3$) [4] and tungsten oxide (WO$_3$) [5] have been used as photoelectrodes. Complex metal oxides, such as BiVO$_4$, are now of interest. Increased efficiencies using photoelectrodes of polycrystalline BiVO$_4$ doped with molybdenum (Mo) [6-9] or tungsten (W) [8,10-12] or co-doped with MO and W [13,14] have been observed. To fully understand the performance of these material a fundamental study of the electrical properties of well characterized single crystal was performed. Initial report of transport measurements [15] is available. In this report, enhanced Hall measurements improve these transport measurements for carrier density and mobility to lower temperatures reported in reference 15.

Experimental methods

Single crystals of BiVO$_4$ were synthesis from ceramic powders. Complete details of crystal synthesis, characterization and sample preparation are given in reference 15. The sample geometry was rectangular and resistivity and Hall measurements used the van der Pauw method. The sample was oriented such that the measurements were made along the principal crystallographic axis. Ohmic contacts were made using In-Ga eutectic and silver paste.

AC field Hall measurements, while not new, is not a common method. A brief review of the method will help explain the advantage of AC field Hall for measurements of materials like $BiVO_4$. Complete details can be found elsewhere [16]. There is a very well developed methodology for measuring the Hall effect and resistivity using DC fields[17]. The methodology is designed to remove the unwanted effects from the measured voltage. The following sections provide a brief summary of this methodology. These explanations are based on the definitions provided here.

The Hall voltage is proportional to the magnetic field (B), current (I), and Hall coefficient (R_H) and depends inversely on the thickness (t). In an ideal geometry the measured Hall voltage is zero with zero applied field. However, the voltage measured in a practical experiment (V_m) also includes a misalignment voltage (V_o) and a thermal electric voltage (V_{TE}). The misalignment voltage is proportional to the resistivity of the material (ρ), the current and a factor (α) that depends on the geometry. This factor converts resistivity to resistance between the two Hall voltage probes. The thermal electric voltage arises from contacts between two different materials and is independent of the current. The thermal electric voltage does depend on the thermal gradients present.

$$V_m = \frac{R_H i B}{t} + V_o + V_{TE}$$

$$V_m = \frac{R_H i B}{t} + \alpha \frac{\rho}{t} i + V_{TE}$$

The mobility (μ) is the Hall coefficient divided by the resistivity.

$$V_m = \frac{\rho i}{t}(\mu B + \alpha) + V_{TE}$$

The factor α can be as small as zero (for no offset), but typically it is about 1.

Current reversal can be used to remove the unwanted effects of thermal electric voltage (V_{TE}). Thermal electric voltage does not depend on current; current reversal exploits this characteristic to remove the effects of V_{TE}. An example of this is shown by measuring the resistance (R) of a sample. The measured voltage, with applied current I_1, can be written as $V_m(I_1) = I_1 R + V_{TE}$. If a second measurement with current I_2 is made, the measured voltage is $V_m(I_2) = I_2 R + V_{TE}$. Normally $I_2 = -I_1$, but allowance is made for cases where I_2 is different from I_1. Then the resistance R can be calculated as $R = (V_m(I_1) - V_m(I_2))/(I_1 - I_2)$. The effect of the thermal electric voltage is removed by subtracting the measured voltage at two different currents.

Field reversal can be used to remove the unwanted effects of the misalignment voltage. Hall voltage depends on the magnetic field, but the misalignment voltage does not depend on the magnetic field. Assuming that the thermal electric voltages have been removed by current reversal, the measured voltage at a field B_1 can be calculated as $V_m(B_1) = \rho I \mu B_1/t + \rho I \alpha/t$ and the measured voltage at a second field B_2 can be calculated as $V_m(B_2) = \rho I \mu B_2/t + \rho I \alpha/t$. As in current reversal, normally $B_2 = -B_1$. Then the quantity $\rho I \mu/t$ is calculated as $\rho I \mu/t = (V_m(B_1) - V_m(B_2))/(B_1 - B_2)$. Since I and t are known quantities, the Hall coefficient ($R_H = \rho \mu$) can be obtained. The Hall resistance is defined as $\rho \mu * (B_1 - B_2) = (V_m(B_1) - V_m(B_2))/I$

For low mobility materials, the quantity μB can be very small compared to α. When the expression $(V_m(B_1) - V_m(B_2))$ is calculated, the subtraction between the two large numbers gives a small result. Any noise in the measurement can easily dominate the actual quantity, and consequently, produce imprecise results. This is often the reason that Hall measurements on low mobility materials give inconsistent carrier signs.

Decreasing the thickness of the sample, or increasing the current, does not necessarily improve the measurements. Both the Hall voltage and the misalignment voltage increase by the same amount and does not change the size of μB relative to α.

A second problem is that the two measurements $V_m(B_1)$ and $V_m(B_2)$ can be separated in time by a significant amount. The time to reverse the field of a magnet can vary from seconds to minutes depending on the magnet configuration. The misalignment voltage $V_0 = \rho I\alpha/t$ depends on the resistivity of the material. If the material changes temperature between the two measurements $V_m(B_1)$ and $V_m(B_2)$, the misalignment voltage will change, and the subtraction will not cancel the misalignment voltage. The un-canceled misalignment voltage will be included in the calculation of the Hall coefficient.

A second method to remove the effect of the misalignment is to use an AC magnetic field. If the magnetic field is made a sinusoidal signal ($B(t) = B \cos(\omega t)$), then in the quasi-static approximation, the Hall voltage will become time dependent as well, $V_H(t) = i\,\rho\mu/t\, B\cos(\omega t)$. The misalignment voltage is independent of the magnetic field, and consequently remains a DC voltage. The measured voltage is now

$$V_m = \frac{\rho i}{t}(\mu B \cos(\omega t) + \alpha)$$

The measurement electronics using a lock-in amplifier can separate the desired AC signal from the undesired DC signal with a high degree of precision. However, there is a new term in the measured voltage. This is proportional to the time derivative of the magnetic field, and it is proportional to the inductance of the sample and the leads used in the measurement. If the proportionality constant is β, the measured voltage should be written as

$$V_m = \frac{\rho i}{t}(\mu B \cos(\omega t) + \alpha) + \beta \frac{dB}{dt}$$

Since this is an AC signal, the lock-in will measure this term as well as the Hall voltage term. Since this term is independent of the current, just like the thermal electric voltage, one method is to use current reversal to remove this term. This term is also 90° out of phase from the signal. Phase resolution on the lock-in amplifier can eliminate this term using a combination of current reversal and phase measurement.

DISCUSSION

Electronic Transport

Resistivity was measured from 300 to 175 K, increasing by 4 orders of magnitude as the temperature was decreased (Figure 1). In many metal oxides, carrier transport is described by a thermally activated small polaron hopping (SPH) mechanism first proposed by Mott.[18] In this model, the charge carrier distorts the surrounding lattice, impeding its transport and resulting in low mobility: an upper limit of $0.1-1$ cm^2/(Vs) has been calculated.[19] The small polaron model is described by [20]

$$\rho(T) \propto T \exp\left(\frac{E_a}{k_B T}\right)$$

where E_a is the hopping activation energy, k_B is Boltzmann's constant, and T is the absolute temperature.

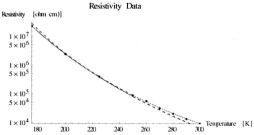

Figure 1. Resistivity of BiVO4 vs. temperature. Plots are fits to different theories in the temperature ranges (see below). The red line for T> 250K is a fit to small polaron hopping model. The blue line for T < 250 is a fit to variable range hopping theory. The dotted lines are extensions of the theory beyond the fit range (see text)

As shown in Figure 1,(red line) this model fits the data closely from 250 to 300 K, and activation energies of 0.271eV were determined. A transition to a variable range hopping (VRH) mechanism is expected at low temperatures, estimated to be 250K for BiVO$_4$ [15]. The Efros-Shkolvskii model of VRH is used to fit the data below 250K. With this model the resistivity is

$$\rho(T) \propto \exp\left(\frac{1}{T^{1/2}}\right)$$

The blue line in Figure 1 is a fit to the resistivity data in the region from 175K to 250K. The dotted blue and red lines in figure 1 are the extension of the small polaron hopping model (red line) and variable range hopping model beyond the fitting range. There is a small but consistent preference for two different hopping models. A more detail discussion on this point can be found in ref 15.

Measurements of the Hall voltage on this material, using DC fields, did not produce consistent results, for reason explained above. [15] Using AC field Hall method the Hall voltage was initially measured from 250K to 300K. The density (n), derived from the Hall voltage as

$$n = \frac{I\,B}{V_H\,e}$$

Here I is the current use for the measurement (100 na for the range 250K to 300K), B is the magnitude of the applied AC field (0.61 T), V_H is the magnitude of the measured Hall voltage and e is the charge of the electron. Figure 2 shows the sheet carrier density vs. temperature in this temperature range. The Hall mobility of BiVO$_4$ is calculated from the resistivity data and the Hall coefficient. Figure 3 is the mobility data for the temperature range 250 K to 300 K.

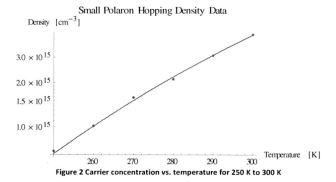

Figure 2 Carrier concentration vs. temperature for 250 K to 300 K

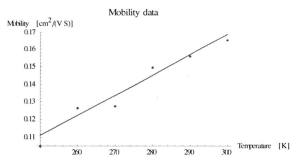

Figure 3 Hall mobility vs. temperature for 250 K to 300 K

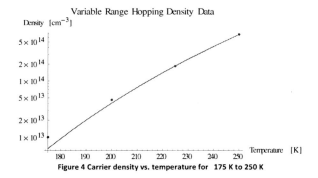

Figure 4 Carrier density vs. temperature for 175 K to 250 K

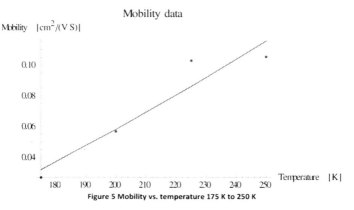

Figure 5 Mobility vs. temperature 175 K to 250 K

Although this data was measured on a different sample than presented in reference [15], the results are consistent with the data presented in [15]. In reference [15], the ac field Hall measurement was limited to this temperature range because of noise in the measurement. To reduce this noise the filtering used in the measurement was improved and additional Hall measurements were possible to 175 K. Below this temperature the resistance of the sample was so high the capacitance effects in cables limited the AC measurement [16]. Figure 1 shows the resistivity measurement in the temperature range of 175K to 250K. Figure 4 is the density and figure 5 is the Hall mobility in this temperature range.

CONCLUSIONS

Hall measurements using AC field and enhanced filtering methods extend the range of measurements of single crystal $BiVO_4$ to 175 kelvin and mobility of 0.03 cm^2/(V s). In the temperature range of 175 K to 200 K variable range hopping describes the electronic transport. In the range of 250 K to 300 K small polaron hopping theory describes the electronic transport.

REFERENCES

(1) Bard, A. J.; Fox, M. A. Acc. Chem. Res. 1995, 28, 141.
(2) Grätzel, M. Nature 2001, 414, 338.
(3) Ni, M.; Leung, M. K.; Leung, D. Y.; Sumathy, K. Renewable Sustainable Energy Rev. 2007, 11, 401.
(4) Sivula, K.; Le Formal, F.; Grätzel, M. ChemSusChem 2011, 4, 432.
(5) Liu, X.; Wang, F.; Wang, Q. Phys. Chem. Chem. Phys. 2012, 14, 7894.
(6) Pilli, S. K.; Furtak, T. E.; Brown, L. D.; Deutsch, T. G.; Turner, J. A.; Herring, A. M. Energy Environ. Sci. 2011, 4, 5028.

(7) Luo, W.; Yang, Z.; Li, Z.; Zhang, J.; Liu, J.; Zhao, Z.; Wang, Z.; Yan, S.; Yu, T.; Zou, Z. Energy Environ. Sci. 2011, 4, 4046.
(8) Luo, W.; Wang, J.; Zhao, X.; Zhao, Z.; Li, Z.; Zou, Z. Phys. Chem. Chem. Phys. 2013, 15, 1006.
(9) Luo, W.; Li, Z.; Yu, T.; Zou, Z. J. Phys. Chem. C 2012, 116, 5076.
(10) Ye, H.; Park, H. S.; Bard, A. J. J. Phys. Chem. C 2011, 115, 12464.
(11) Zhong, D. K.; Choi, S.; Gamelin, D. R. J. Am. Chem. Soc. 2011, 133, 18370.
(12) Abdi, F. F.; Firet, N.; van de Krol, R. ChemCatChem 2013, 5, 490.
(13) Park, H. S.; Kweon, K. E.; Ye, H.; Paek, E.; Hwang, G. S.; Bard, A. J. J. Phys. Chem. C 2011, 115, 17870.
(14) Berglund, S. P.; Rettie, A. J. E.; Hoang, S.; Mullins, C. B. Phys. Chem. Chem. Phys. 2012, 14, 7065.
(15) Rettie, A. J. E., et al. J. Am. Chem. Soc. 2013 **135**(30): 11389-11396.
(16) Lindemuth, J.; Mizuta, S.-I. In SPIE Solar Energy + Technology; International Society for Optics and Photonics: Bellingham, WA, 2011; p 81100I.
(17) Look, D. C. **Electrical characterization of GaAs materials and devices**; Wiley: New York, 1989.
(18) Mott, N. F.; Davis, E. A. **Electronic processes in non-crystalline materials**, 2nd ed.; OUP: Oxford, U.K., 1979.
(19) Emin, D.; Seager, C.; Quinn, R. K. Phys. Rev. Lett. 1972, 28, 813.
(20) Austin, I. G.; Mott, N. F. Adv. Phys. 1969, 18, 41.

A DLTS study of a ZnO microwire, a thin film and bulk material

Florian Schmidt[1], Peter Schlupp[1], Stefan Müller[1], Christof Peter Dietrich[1], Holger von Wenckstern[1], Marius Grundmann[1], Robert Heinhold[2], Hyung-Suk Kim[2], and Martin Ward Allen[2]

[1]Universität Leipzig, Institut für Experimentelle Physik II, Linnéstraße 5, 04103 Leipzig, Germany

[2]The MacDiarmid Institute for Advanced Materials and Nanotechnology, University of Canterbury, Christchurch, New Zealand

ABSTRACT

We have investigated the electrical properties of a ZnO microwire grown by carbo-thermal evaporation, a ZnO thin film grown by pulsed-laser deposition on an a-plane sapphire, and a hydrothermally grown Zn-face ZnO single crystal (Tokyo Denpa Co. Ltd.). The samples were investigated by means of current-voltage measurements, capacitance-voltage measurements, and deep-level transient spectroscopy.

The defects T2 [1,2] and E3 [1,3,4] were identified in all three sample types. Additionally, in the single crystal and thin film samples E64 [5] and E4 [1] were detected. These findings support the common opinion that T2 is an intrinsic defect since it is found in all the samples investigated and thus its occurrence is not related to any growth technique.

INTRODUCTION

ZnO ist the wide band-gap semiconducting oxide that was most intensively studied in the last decade. Nevertheless, the identification of other than effective mass defects is not accomplished yet despite enormous recent efforts. These include besides first-principles investigations, magnetic resonance studies also investigations by space-charge spectroscopic methods, especially deep-level transient spectroscopy (DLTS). Here, the defect parameters thermal activation energy and capture cross-section being a defect's fingerprint are accessible. However, DLTS is "chemically blind" and hence an attribution of an electronic state to a certain impurity is indirectly possible, only. In this letter we investigate the relative concentration of the deep level defects T2 and E3 in ZnO samples that were realized by three different growth methods in order to come closer to their microscopic origin.

EXPERIMENT
Contact Properties
Room temperature current-voltage (IV) measurements were applied using an Agilent 4156C precision semiconductor parameter analyzer and are shown in Fig. 1 (a) for all three ZnO structures.

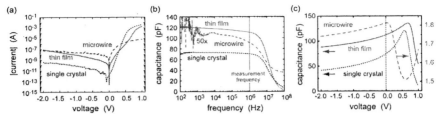

Figure 1. (a) Room temperature current-voltage characteristic of the microwire, thin film and single crystal, respectively. (b) Frequency-and (c) voltage dependence of the capacitance of all samples investigated.

The series resistance R_s, the shunt resistance R_p, and the rectification ratio RR defined by $I(+1\text{ V})/I(-1\text{ V})$ were determined and are summarized in Tab. 1.

Table 1. Series resistance R_s, shunt resistance R_p and rectification ratio RR obtained from room-temperature IV-characteristics of ZnO Schottky diodes.

Sample	R_s (Ω)	R_p (Ω)	RR
Microwire	80k	40M	3×10^2
thin film	61	6M	2×10^5
single crystal	115	9G	6×10^7

Frequency-dependent capacitance measurements reveal the influence of the series resistance on the cut-off frequency ν_{co} and on the frequency dependence of the capacitance. Since ν_{co} is inversely proportional to R_s and this value is quite high for the microwire (80 kΩ, cmp. Tab. 1), it was necessary to check that ν_{co} was above the 1 MHz measuring frequency of our deep-level transient spectroscopy (DLTS) system. This was confirmed for all diodes used in this study, as shown in Fig. 1 (b).

Figure 1 (c) shows the capacitance-voltage (CV) measurements of all samples probed at 1 MHz. All samples show decent capacitance and are therefore suitable for investigations by DLTS.

DLTS studies

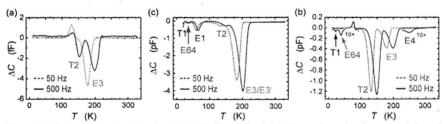

Figure 2. DLTS scans of (a) the microwire, (b) the thin film and (c) the single crystal sample using a rate window of 500 Hz and 50 Hz, respectively.

DLTS was applied in a temperature range from 10 K to 330 K. A description of the DLTS system can be found in Ref. [6]. The samples were biased at $V=-2$ V and excited using a filling pulse of 2.5 V applied for t_p=1 ms. Rate windows in the range of 2.5 Hz to 2000 Hz were used. The DLTS scans of the microwire, the thin film and the single crystal sample are shown in Fig. 2 (a) to (c) for rate windows of 50 Hz and 500 Hz, respectively.

For the ZnO microwire, we found DLTS peaks that correspond to the carrier emission of the defects T2 and E3. The E3 defect, which is commonly observed in ZnO, was also present in the thin film and single crystal samples (cmp. Fig. 2 (b) and (c)). The defect activation energy of approximately 300 meV was also determined by temperature dependent Hall-effect measurements on an undoped ZnO microwire [7].

It was shown that the activation energy and apparent capture cross section σ_n of T2 depends on its concentration. The trap parameters of T2 in the microwire sample would therefore allow an estimation of the defect concentration $N_{t,T2}$. By extrapolation of the σ_n versus $N_{t,T2}$ relation (inset in Fig. 2 in Ref. [2]) we determined the concentration to be below 10^{14} cm^{-3}. In general the appearance of E3 and T2 in the microwire sample is a strong indication for their connection to intrinsic defects.

The concentration of T2 is enhanced in ZnO under annealing in vacuum and an oxygen ambient as shown in melt grown single crystal ZnO and PLD grown ZnO, respectively. While this is confirmed in the low-Li bulk sample prepared at nominally 1400°C ($N_{t,T2}$=3×10^{15} cm^{-3}) as well as in the PLD thin film grown at 650°C ($N_{t,T2}$=4×10^{14} cm^{-3}), surprisingly the concentration of T2 in the microwire is lowest (since E_t is highest) even at intermediate growth temperatures of 950°C. We therefore suggest that the position of the Fermi level E_F significantly affects the concentration of T2. Based on a model introduced by Schmidt et al., the T2 defect is assumed to be a donor-acceptor complex. Although our experimental methods are not sensitive to the chemical nature of either the donor or the acceptor, we would propose the zinc vacancy V_{Zn} as a possible candidate for an acceptor. First-principles studies reveal that V_{Zn} has a lower formation energy for a higher Fermi level [8]. Due to the naturally high concentration of donors in ZnO it is reasonable that the T2 concentration is determined by the concentration of that acceptor and with that the position of the Fermi level. This is supported by the net-doping density N_d-N_a, which is a measure for E_F in our samples. N_d-N_a was obtained from capacitance-voltage measurements for the PLD thin film and the low-Li bulk sample to be 1×10^{17} cm^{-3} and 4×10^{16} cm^{-3}, respectively.

Due to the large uncertainty of the effective contact area we can only estimate the net doping density for the microwire to be about 10^{16} cm^{-3}. This value is consistent with the apparent free carrier concentration n_H=2×10^{16} cm^{-3} determined from Hall effect measurements of similar wires. Therefore, we conclude that the acceptor is most likely formed in the PLD thin film followed by the bulk sample and the microwire, which is consistent with the model proposed for T2.

A positive DLTS signal occurs at approximately 130 K for the microwire sample and at approximately 75 K in the single crystal. The peak does not shift in temperature but in amplitude by applying different rate windows, therefore this signal is most likely due to the carrier freeze-out. This leads to a strong change of the capacitance as well as the series resistance. The latter explains the sign-change of the DLTS signal as shown by Broniatowski et al. [9]. No positive DLTS signal was detected in the thin film sample where the freeze-out occurs at much lower temperatures. Shallow defects below 100 K (e.g. T1, E64, E1), which are detectable in the thin film and in the single crystal, are not measurable by DLTS, which is again in agreement with the low Fermi energy in this sample. However, low temperature cathodoluminescence measurements showed that for instance the Al-donor Al_{Zn} (E64) is also incorporated in ZnO microwires [5].

Freeze-out due to the comparatively large degree of compensation in the microwire sample and the as-grown ZnO-bulk sample supports the assumption that V_{Zn} is preferably formed in the PLD thin film having a high Fermi level, while the generation of the defect is less distinct in the compensated microwire or only occurs after thermal treatment as it is the case for the two bulk samples.

CONCLUSIONS

In this work, we present a comparative study of defects in differently grown ZnO material, including a microwire grown by carbo-thermal evaporation, a pulsed laser-deposited thin film and a hydrothermally grown bulk crystal. IV and CV measurements were performed in order to study the electric properties of the samples.
In the ZnO microwire the deep defects T2 and E3 were detected by means of deep-level transient spectroscopy. The absence of other deep defects confirms the high quality of such ZnO material. Since T2 and E3 are found in all ZnO samples investigated it is most likely that they are of intrinsic origin. Complete freeze-out of carriers at approximately 130K screens shallow defect levels in the microwire sample.

ACKNOWLEDGMENTS

We thank Gabriele Ramm, Holger Hochmuth, and Michael Lorenz for preparing the PLD targets and growing the samples. To Monika Hahn, who prepared the samples, we are much obliged. This work was financially supported by the German Science Foundation (DFG) in the framework of SFB 762 (Functionality of Oxide Interfaces) and the Graduate School Leipzig School of Natural Sciences – BuildMoNa (GS185/1).

REFERENCES

1. T. Frank, G. Pensl, R. Tena-Zaera, J. Zúñiga-Pérez, C. Martínez-Tomás, V. Muñoz-Sanjosé, T. Ohshima, H. Itoh, D. Hofmann, D. Pfisterer, J. Sann, B. Meyer, Appl. Phys. A 88, 141 (2007).
2. M. Schmidt, M. Ellguth, R. Karsthof, H. von Wenckstern, R. Pickenhain, M. Grundmann, G. Brauer, F. C. C. Ling, Phys. Status Solidi B 249, No. 3, 588 (2012).
3. H. von Wenckstern, M. Brandt, H. Schmidt, G. Biehne, R. Pickenhain, H. Hochmuth, M. Lorenz, M. Grundmann, Appl. Phys. A 88, 135 (2007).
4. A. Rohatgi, S. K. Pang, T. K. Gupta, and W. D. Straub, J. Appl. Phys. 63, 5375 (1988).
5. H. von Wenckstern, G. Biehne, M. Lorenz, M. Grundmann, F. D. Auret, W. E. Meyer, P. J. J. van Rensburg, M. Hayes, and J. M. Nel, J. Korean Phys. Soc. 53, 2861 (2008).
6. F. Schmidt, H. von Wenckstern, D. Spemann, M. Grundmann, Appl. Phys. Lett. 101, 012103 (2012).
7. C. P. Dietrich, M. Brandt, M. Lange, J. Kupper, T. Böntgen, H. von Wenckstern, and M. Grundmann, J. Appl. Phys. 109, 013712 (2011).
8. F. Oba, M. Choi, A. Togo, and I. Tanaka, Sci. Technol. Adv. Mater. 12, 034302 (2011).
9. A. Broniatowski, A. Blosse, P. C. Srivastava, and J. C. Bourgoin, J. Appl. Phys. 54, 2907 (1983).

Evaluation of Sub-Gap States in Amorphous In-Ga-Zn-O Thin Films Treated with Various Process Conditions

Kazushi Hayashi, Aya Hino, Hiroaki Tao, Yasuyuki Takanashi, Shinya Morita, Hiroshi Goto, and Toshihiro Kugimiya
Electronics Research Laboratory, Kobe Steel, Ltd., 1-5-5 Takatsuka-dai, Nishi-ku, Kobe, 651-2271, Japan.

ABSTRACT

In the present study, the sub-gap states of amorphous In-Ga-Zn-O (a-IGZO) thin films treated with various process conditions have been evaluated by means of capacitance-voltage (C-V) characteristics and isothermal capacitance transient spectroscopy (ICTS). It was found that the space-charge densities of the a-IGZO decreased as the oxygen partial pressure was increased during the sputtering of a-IGZO thin films. The ICTS spectra for the 4, 8, and 12 % samples were similar and the peak positions were found to be around 1×10^{-2} s at 180 K. On the other hand, the peak position for the 20 % sample shifted to a longer time regime and was located at around 2×10^{-1} s at 180 K. The total densities of the traps for the 4, 8, and 12 % samples were calculated to be $5-6 \times 10^{16}$ cm^{-3}, while that for 20 % was one order of magnitude lower than the others. From Thermal desorption spectrometer, it was found that desorption of Zn atoms started at a temperature higher than 300 °C for the 4 % sample, while desorption of Zn was not observed for the 20 % sample. The introduction of the sub-gap states could be attributed to oxygen-rich and/or Zn-deficient defects in the a-IGZO thin films formed during thermal annealing.

INTRODUCTION

It is widely recognized that oxide semiconductors such as amorphous In-Ga-Zn-O (a-IGZO) are promising channel materials for thin-film transistors (TFTs) in next generation flat panel display since they have a higher electron mobility than conventional materials like amorphous Si [1,2]. A crucial issue for realization of practical TFTs is comprehension of their electronic properties, information especially regarding sub-gap states in the a-IGZO thin films that govern the TFT performance. The nature of the sub-gap states such as their density and activation energy should be clarified since it is varied depending on the fabrication process and parameters.

Isothermal capacitance transient spectroscopy (ICTS), in which the capacitance transient response originated from an emission of captured carriers is measured under isothermal conditions, is a highly sensitive and spectroscopic technique to directly determine the energy level and the density of carrier traps [3-5]. In the present study, capacitance-voltage (C-V) characteristics and ICTS measurements were applied to evaluate the sub-gap states in the a-IGZO thin films treated with various processes. The influence of the oxygen partial pressures during the a-IGZO sputtering on the electronic properties of the a-IGZO thin films was discussed.

EXPERIMENTAL

A series of metal oxide semiconductor (MOS) diodes with bottom gate structure was fabricated for evaluation. Mo thin films that act as gate electrodes were deposited on glass substrates by DC sputtering. Then, gate insulators (G/Is) consisting of SiO_2 were grown by plasma-enhanced chemical vapor deposition with a mixture of SiH_4 and N_2O diluted by N_2. The thickness of the G/Is was 250 nm. As an active layer, a-IGZO thin films with a thickness of 500 nm were deposited by DC sputtering in a mixture of Ar/O_2 at room temperature. A standard sputtering target with a nominal composition of In:Ga:Zn = 1:1:1 was used for the deposition. The input power for plasma excitation was 200 W and the process pressure was 1×10^{-3} Torr. The oxygen partial pressures during the a-IGZO sputtering were set at 0, 4, 8, 12, and 20 % by controlling of the gas flow rate.

After patterning the active layer using photolithography, the a-IGZO thin films were annealed at 350 °C for 1 h in air. After the annealing, a 100-nm thick SiO_2 layers were deposited by p-CVD in order to simulate the etch stop layer (ESL) for conventional ESL-type TFTs. After through-hole etching, Mo thin films were deposited by DC sputtering as ohmic electrodes. Finally, SiO_2/SiN stacked layers were deposited by p-CVD for passivation. The diameter of the diodes was 1 mm (the active area $A = 0.7854$ mm^2).

The $C-V$ characteristics were measured at room temperature under dark conditions using LCR meters. The measurement frequency was 1 MHz. The sub-gap states in the a-IGZO thin films were evaluated by ICTS using a DA-1500 (Horiba, Ltd). The pulse width W_p was set at 100 ms to fill up all sub-gap states below the Fermi level with carriers, and the sampling range $T = 1$ s. The measurements were performed in a temperature range between 170 and 210 K.

Thermal desorption spectrometer (TDS) was performed to evaluate the thermal desorption of elements from the as-deposited a-IGZO thin films. The samples were heated at a rate of 60 °C/min. The base pressure before the measurements was below 3.7×10^{-9} Torr. Depth profiles of H and OH were detected by secondary ion mass spectroscopy (SIMS). Cs^+ ions were used as the primary ions.

RESULTS AND DISCUSSION

Figure 1(a) shows the $C-V$ characteristics of the MOS diodes fabricated with various oxygen partial pressures during the a-IGZO sputtering. The observed capacitance for the 4, 8, 12, and 20 % samples increased with the biasing voltages, which confirmed that the depletion layers formed at the interface between the a-IGZO thin films and the G/Is. On the other hand, an apparent capacitance change was not observed for the sample with 0 % oxygen partial pressure. The features were varied depending on the a-IGZO sputtering conditions, which implied there was a variety of the space-charge density. Figure 1(b) shows the distribution of the space-charge density as a function of the distance from the interfaces for the 4, 8, 12, and 20 % samples, estimated from the thickness of the depletion layer and the slope of conventional $1/C^2-V$ curve. It is shown that the distributions of the space charge for all samples were not uniform. The space-charge densities near the surface region for all samples were about 1×10^{18} cm^{-3} and decreased with the distance from the interfaces down to 50 nm. The space-charge densities for the 4 % and 8 % samples gradually increased to around 1×10^{18} cm^{-3}, while the space-charge densities for the 12 and 20 % samples continued to decrease and lowered by 1–2 orders of magnitude than those for the 4 and 8 % samples. Note that the 4 and 8% samples were not completely depleted throughout the films since the space-charge densities for the 4 and 8% samples are high.

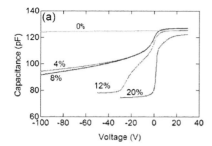 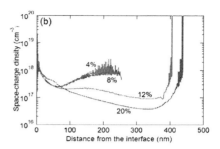

Figure 1. (a) C-V characteristics and (b) distribution of the space-charge densities for the 500-nm MOS diodes with various oxygen partial pressures during the a-IGZO sputtering.

The carrier concentrations for the 0, 4, 8, and 12 % samples, measured by Hall effect measurements, were 1.3×10^{19}, 5.4×10^{16}, 2.9×10^{16}, and 1.7×10^{15} cm^{-3}, while the resistivity of the films were 2.7×10^{-2}, 2.7, 5.2, and 1.1×10^{2} Ω cm. The resistivity of the 20 % sample was higher than the detection limit of the system (over 10^4 Ω cm) and suggested that the carrier concentration of the films was below 10^{14} cm^{-3}. The features are consistent with the general observation that the resistivity of the films increased with the oxygen partial pressure during the deposition [6-8].
Figure 2 shows the ICTS spectra of the a-IGZO MOS diodes with the oxygen partial pressures of (a) 4 %, (b) 8 %, (c) 12 %, and (d) 20 %. The measurements were performed in a temperature range between 170 and 210 K. The conditions for the 4, 8, and 12 % samples were the biasing voltage = -1 V and the pulse height = 6 V, while that for the 20 % sample were 0 V and 5 V, respectively. The broad peaks that shifted toward a shorter time regime with measurement temperature were clearly observed from all samples. The features for the 4, 8, and 12 % samples were similar and the peak positions were around 1×10^{-2} s at 180 K. On the other hand, the peak position for the 20 % sample shifted to a longer time regime and was around 2×10^{-1} s at 180 K, which implied that the a-IGZO with 20 % oxygen partial pressure possessed the slightly deeper sub-gap states below the conduction band minimum (CBM).
In general, the activation energy (ΔE_T) and the capture cross section of the traps (σ_T) can be correlated with the following equation: [3]

$$\frac{1}{\tau} = N_C \sigma_n v_{th} \cdot \exp\left(-\frac{\Delta E}{kT}\right)$$

where τ is time constant of the ICTS spectra, N_C is effective density of states in the CBM, σ_n is capture cross section of electrons, and v_{th} is the thermal velocity. The activation energies of the sub-gap states estimated from the results shown in Figs. 2 were roughly estimated to be around 0.2 eV below the CBM although the obtained peaks were broader than those predicted by the theory of the ICTS and detailed determination proved to be difficult due to overlapping of the spectra.

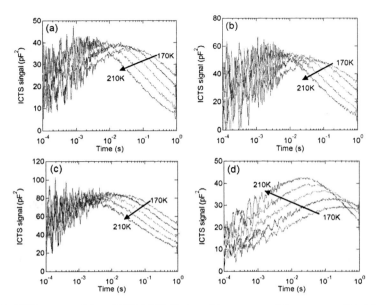

Figure 2. ICTS spectra of the a-IGZO MOS diodes with oxygen partial pressures of (a) 4 %, (b) 8 %, (c) 12%, and (d) 20%. The measurement temperatures were 170, 180, 190, 200, and 210 K.

The gap states at around 0.2 eV below the CBM are often observed in ZnO based materials [4,5,9-11]. Indeed, it has been reported that the localized states exist at 0.1–0.3 eV below the CBM in the a-IGZO thin films from the C-V [11] and ICTS [4, 5] measurements.

Since the ICTS spectra suggested that the activation energy of the sub-gap states distributed continuously, the total density of the traps (N_T) for each sample was compared from differences in the square of C between $t = 1 \times 10^{-5}$ to 1 s in the C-t curve. It was calculated that N_T for the 4, 8, and 12 % samples were about 5–6 × 10^{16} cm^{-3}, while N_T for 20 % was found to be one order of magnitude lower than the others.

It is known that the properties of the a-IGZO thin films are markedly changed by thermal annealing. Therefore, TDS was performed. Figures 3 shows the results of TDS from a-IGZO film for the 4 and 20 % samples. It can be seen that desorption of O_2 started at around 100 °C for both samples, but desorption tends to be suppressed as the measurement temperature is increased for the 20 % sample. On the other hand, desorption of Zn atoms started at a temperature higher than 300 °C for the 4 % sample, while desorption of Zn was not observed for the 20 % sample.

Recently, Quemener et al., reported that, based on their DLTS measurements from ZnO with various annealing treatments, the deep level at around 0.2 eV (known as E2) originated from oxygen-rich and/or Zn-deficient defects [9]. Considering the fact that the 0.2 eV gap states were not observed in the 20 % sample, in addition to the origin of the defect states in ZnO, the introduction of the gap states can be correlated with the behavior of thermal desorption of Zn atoms in the a-IGZO thin films. Furthermore, the increased level of desorption of O_2 in the 4 % sample implies that there is formation of oxygen deficiency in the film.

Figures 4(a) and (b) show the H and OH distributions in the a-IGZO thin films detected by SIMS. Incorporation of hydrogen atoms was confirmed. Nomura *et al.*, pointed out that the hydrogen atoms existed in the form of hydroxyl group –OH, but most of them were inactive [12]. As clearly shown in Fig. 4(c), the distributions of H for the 4 and 12 % were different from that for OH; they gradually increased toward the ohmic contacts (Mo). It is likely that the increase in the space charge originated from the introduction of the oxygen deficiency, which is reported to be responsible for the shallow donor states in ZnO based semiconductors, forming complex with hydrogen atoms [13].

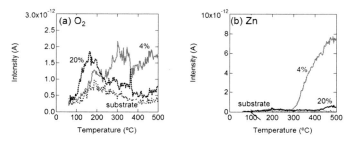

Figure 3. Results of TDS from a-IGZO film for 4 % and 20 % samples. (a) for O_2 and (b) for Zn.

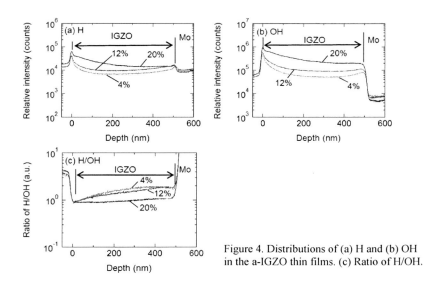

Figure 4. Distributions of (a) H and (b) OH in the a-IGZO thin films. (c) Ratio of H/OH.

CONCLUSIONS

The a-IGZO thin films with various oxygen partial pressures were deposited and evaluated by the C-V and the ICTS measurements. The space-charge densities of the a-IGZO decreased as the oxygen partial pressure increased during the sputtering of a-IGZO thin film. The ICTS measurements revealed that the peak positions for the 4, 8, and 12 % samples were around 1×10^{-2} s at 180 K, while that for the 20 % sample shifted to a longer time regime and was at around 2×10^{-1} s at 180 K. The values of N_T for the 4, 8, and 12% samples were 5–6×10^{16} cm^{-3}, which was one order of magnitude higher than that for 20 %. From TDS, it was found that desorption of Zn atoms started at a temperature higher than 300 °C for the 4 % sample, while desorption of Zn was not observed for the 20 % sample. The introduction of the sub-gap states could be attributed to the oxygen-rich and/or Zn-deficient defects in the a-IGZO thin films formed during the thermal annealing.

ACKNOWLEDGMENTS

The authors would like to thank Dr. H. Okada of KOBELCO Research Institute, Inc., for his helpful discussion on the ICTS measurements. Technical support of M. Urano of ESCO Ltd. is also acknowledged.

REFERENCES

1. K. Nomura, H. Ohta, A. Takagi, T. Kamiya, M. Hirano, and H. Hosono, Nature 488, 432 (2004).
2. T. Kamiya, K. Nomura, and H. Hosono, Sci. Technol. Adv. Mater. **11**, 044305 (2010).
3. H. Okushi and Y. Tokumaru, Jpn. J. Appl. Phys. **19**, L335 (1980).
4. K. Hayashi, A. Hino, S. Morita, S. Yasuno, H. Okada, and T. Kugimiya, Appl. Phys. Lett. **100**, 102106 (2012).
5. A. Hino, S. Morita, S. Yasuno, T. Kishi, K. Hayashi, and T. Kugimiya, J. Appl. Phys. 112, 114515 (2012).
6. H. Yabuta, M. Sano, K. Abe, T. Aiba, T. Den, and H. Kumomi, Appl. Phy. Lett. 89, 112123 (2006)
7. D. Kang, H. Lim, C. Kim, I. Song, J. Park, Y. Park, and J. Chung, Appl. Phy. Lett. 90, 192101 (2007).
8. T. Kamiya and H. Hosono, ECS Transactions, 54 (1) 103 (2013).
9. V. Quemener, L. Vines, E. V. Monakhov, and B. G. Svensson, Appl. Phy. Lett. 100, 112108 (2012).
10. W. Mtangi, F. D. Auret, W. E. Meyer, M. L. Legodi, P. J. Janse van Rensburg, S. M. M. Coelho, M. Diale, and J. M. Nel, J. Appl. Phys. 111, 094504 (2012).
11. M. Kimura, T. Nakanishi, K. Nomura, T. Kamiya, and H. Hosono, Appl. Phys. Lett. 92, 133512 (2008).
12. A. Janotti and C. G. Van de Walle, Rep. Prog. Phys. 72, 126501 (2009).
13. K. Nomura, T. Kamiya, and H. Hosono, ECS J.Solid State Sci.Tech. 2, P5 (2013).

Effects of N_2O addition on the properties of ZnO thin films grown using high-temperature H_2O generated by catalytic reaction

Naoya Yamaguchi[1], Eichi Nagatomi[1], Takahiro Kato[1], Koichiro Ohishi[2], Yasuhiro Tamayama[1], and Kanji Yasui[1]
[1] Nagaoka University of Technology, Nagaoka, Niigata 940-2188, Japan
[2] Nagaoka National College of Technology, Nagaoka, Niigata 940-8532, Japan

ABSTRACT

The effects of N_2O gas addition on the properties of zinc oxide films grown on a-plane (11-20) sapphire (a-Al_2O_3) substrates were investigated, using a chemical vapor deposition method based on the reaction between dimethylzinc and high-energy H_2O produced by a Pt-catalyzed H_2-O_2 reaction. By employing an optimal N_2O gas pressure, both the film crystallinity and crystal orientation were improved. Subsequent to treatment with N_2O, the electron mobility of films at room temperature increased from 207 to 234 cm^2/Vs while the electron concentration decreased at low temperatures. In addition, the photoluminescence peak intensity of the near-band-edge emission was increased.

INTRODUCTION

Zinc oxide (ZnO) is a useful material in many applications, such as surface acoustic wave devices [1], gas sensors [2], photoconductive devices [3] and transparent electrodes [4]. Due to its large bandgap (3.37 eV at RT) and large exciton binding energy (60 meV) [5], its application to optoelectronic devices such as light emitting diodes and laser diodes operating in the ultraviolet region has been intensively investigated [6-16]. Many growth techniques, including molecular beam epitaxy (MBE) [6-9], pulsed laser deposition (PLD) [10-12], laser MBE (LMBE) [13], atomic layer deposition [14] and metal-organic chemical vapor deposition (MOCVD) [15-18], have been used to prepare ZnO thin films. Although MOCVD has many advantages in industrial processes, such as a high growth rate on large surface substrates and a wide selection of metal-organic and oxygen source gases, ZnO film growth by conventional MOCVD requires the application of a large amount of electrical energy to react the source gases and raise the substrate temperature. To eliminate this drawback, a more efficient means of reacting oxygen and metal-organic source gases is required. When thermally-excited water is used to hydrolyze the metal-organic source gases, high-energy ZnO precursors are produced in the gas phase, thus allowing the growth of ZnO films in a manner similar to PLD and MBE. In a previous paper [19], we reported a new technique for growing ZnO films using the reaction between dimethylzinc (DMZn) and high-temperature H_2O produced by a Pt-catalyzed H_2-O_2 reaction, and also detailed the excellent electrical and optical properties of ZnO films grown on a-plane (11-20) sapphire (a-Al_2O_3) substrates.

To fabricate light emitting diodes and laser diodes, the growth of p-type ZnO films is required. However, the reproducible growth of p-type ZnO films on sapphire substrates is very difficult and nitrogen oxide gases such as N_2O and NO are often used [20, 21]. In this paper, we report an investigation of the effects of the N_2O gas supply used during film growth on the properties of the resultant ZnO films. The crystallinity, electrical properties and photoluminescence spectra of ZnO films grown with and without N_2O gas are described.

EXPERIMENTAL

The CVD apparatus used in this study was the same as that presented in a previous paper [19]. H_2 and O_2 gases were introduced into a catalyst cell containing Pt-dispersed ZrO_2 catalyst. The temperature of the catalyst cell increased rapidly to over 1000 °C following the introduction of the H_2-O_2 gas supply, due to the exothermic reaction of H_2 and O_2 on the catalyst, although the cell temperature stabilized within five minutes after the introduction of the gases. A shutter placed between a skimmer cone and the substrate holder was opened 10 min after the gases were introduced and the resulting thermally excited H_2O molecules were ejected from a fine nozzle into the reaction zone and allowed to collide with streams of DMZn and N_2O ejected from additional fine nozzles. Although the H_2 and O_2 gas flow rates varied somewhat between deposition trials, typical values were 400 and 130 sccm, respectively. The DMZn gas flow rate was ascertained by monitoring the pressure and had a typical value of 4.0×10^{-3} Pa. The reaction gas pressure in the chamber during the deposition process ranged from 0.4 to 1.0 Pa; this variability resulted from setting the burning temperature in the catalytic cell to the maximum possible values of over 1000 °C during each deposition experiment. The distance between the H_2O nozzle and the substrate was 50 mm. The skimmer cone was placed between the H_2O nozzle and the substrate to select only high-velocity H_2O molecules and direct them to the substrate.

The epitaxial ZnO films (6–8 μm thick) were grown directly on a-Al_2O_3 substrates at a substrate temperature of 773 K for 60 min without a buffer layer. Although the N_2O gas supply pressure to the reaction zone was varied from 3.2×10^{-3} to 9.7×10^{-2} Pa, all films exhibited n-type characteristics. The highest values of electron mobility were obtained upon the addition of N_2O gas at a pressure of 3.2×10^{-3} Pa. Prior to use, sapphire substrates were degreased with methanol and acetone, etched with an H_2SO_4/H_3PO_4 solution, then rinsed with ultrapure water and finally set on a substrate holder in the CVD chamber. The film thickness obtained in the absence of N_2O gas was 8 μm, while films grown under an N_2O pressure of 3.2×10^{-3} Pa were 6 μm thick. The crystallinity and crystal orientation of each film was measured using an x-ray diffractometer with two Ge(220) analyzer crystals (RIGAKU, RINT Ultima III). Hall mobility and residual carrier concentrations were determined using the van der Pauw method (ECOPIA, HMS-5000) with a magnetic field of 0.57 T at room temperature. The temperature dependencies of the Hall mobility and residual carrier concentration were also examined between 80 and 290 K. Photoluminescence (PL) spectra were acquired at room temperature using a spectrometer (HORIBA JOBIN YVON, LabRAM HR-PL) in conjunction with a CCD detector (HORIBA JOVAN YVON, Synapse), applying excitation with the 325 nm line of a He-Cd laser.

RESULTS AND DISCUSSION

Figure 1 shows the X-ray diffraction (XRD) patterns of ZnO films grown without N_2O gas and grown under N_2O at a pressure of 3.2×10^{-3} Pa. The intense (0002) peak associated with the ZnO(0001) index plane can be observed at $2\theta = 34.45$-$34.46°$, which appears at a slightly higher angle than the position of the corresponding peak of the bulk ZnO crystal ($2\theta = 34.41$-$34.42°$).

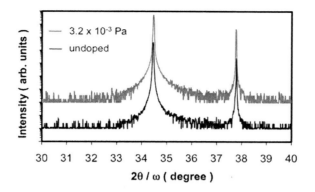

Fig. 1 X-ray diffraction patterns of ZnO films grown with and without N_2O gas

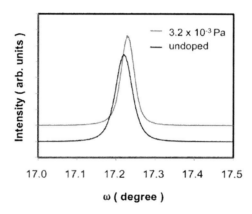

Fig. 2 ω-rocking curves of ZnO films grown with and without N_2O gas

This suggests that the $ZnO/a-Al_2O_3$ films grown using this method contain a small amount of residual tensile stress. In spite of the lower film thickness, the peak intensity of the (0002) diffraction of the ZnO film made using N_2O gas is approximately 1.3 times greater than that of the ZnO film obtained without N_2O. The (11-20) peak associated with $a-Al_2O_3$ is also observed at $2\theta = 37.77°$. The full width at half maximum (FWHM) of the ω-rocking curve of the ZnO (0002) peak grown under N_2O at a pressure of 3.2×10^{-3} Pa is presented in Figure 2. The FWHM of this peak is 147 arcsec, which is less than that of the peak obtained from the film made without N_2O (185 arcsec). From these results, the crystallinity and crystal orientation along the c-axis are evidently improved by growth under a N_2O pressure of 3.2×10^{-3} Pa. The in-plane (φscan) diffraction patterns of the ZnO films produced with and without N_2O gas (the same samples shown in Figures 1 and 2) exhibit intense (10-10) peaks at intervals of 60° (data not shown). No other peaks were evident between the intense peaks, implying the absence of rotational domains in both films.

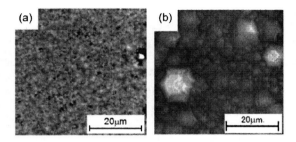

Fig. 3 AFM images for ZnO films grown (a) without and (b) with N_2O gas

Figure 3 shows AFM images of the surfaces of ZnO films produced without and with N_2O (at a pressure of 3.2×10^{-3} Pa). Although hexagonal facets can be observed in both films, the dimensions of the facets are significantly larger in the ZnO film grown with N_2O, in which facets greater than 10 μm in diameter can be observed.

Figure 4 plots the temperature dependencies of the electron mobilities of ZnO films grown with and without N_2O. The electron mobility of the N_2O-doped (3.2×10^{-3} Pa) film at 290 K was 234 cm^2/Vs, while that of the undoped ZnO film was 207 cm^2/Vs. The electron mobilities of both ZnO films increased significantly with decreases in temperature down to 100 K, but subsequently decreased at temperatures less than 100 K. The mobility of the N_2O doped film (234 cm^2/Vs at 290 K) increased to 1100 cm^2/Vs at 100 K.

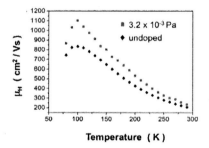

Fig. 4 Temperature dependencies of electron mobilities in ZnO films produced with and without N_2O gas

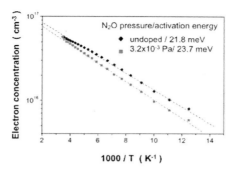

Fig. 5 Temperature dependencies of electron concentrations in ZnO films produced with and without N_2O gas

Figure 5 shows the temperature dependencies of the electron concentrations of the ZnO films. The carrier concentrations of both films were 5.8×10^{16} cm^{-3} at 290 K, although the concentration of the N_2O-doped ZnO film decreased below that of the undoped film as the temperature was lowered from 280 to 80 K. The activation energy of the N_2O-doped ZnO film as obtained from the associated Arrhenius plot was 23.7 meV, while that of the undoped ZnO film was 21.8 meV. Although donor impurities are partly compensated for by the nitrogen acceptor, the intrinsic donor impurities caused by defects appear to be reduced by doping with nitrogen, in view of the significant electron mobility of the N_2O-doped ZnO film.

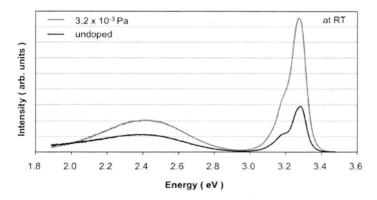

Fig. 6 Room temperature photoluminescence spectra of ZnO films produced with and without N_2O gas

Figure 6 presents the PL spectra of the undoped and N_2O-doped ZnO films, as acquired at room temperature. Both films show two emission bands; the near band-edge emission (NBE) in the ultraviolet region and the broad deep-level emission (DLE) band centered at 2.4-2.5 eV. The origin of the DLE band is attributed to either oxygen vacancy (V_O) or interstitial Zn (Zn_i) defects

[22]. The relative peak intensity ratios of the NBE to DLE bands were 2.6 and 4.2 for the undoped and N_2O doped ZnO films, respectively. Compared to the undoped film, the N_2O doped film exhibited greater NBE intensity and NBE/DLE ratio. Structural defects, such as dislocation and grain boundaries, can trap photogenerated carriers by nonradiative recombination processes before the emission occurs. The addition of N_2O to the ZnO film may eliminate the generation of V_O or Zn_i defects and the increased intensity of the NBE band and the higher NBE/DLE ratio indicate that the N_2O doped ZnO film possesses improved optical quality due to its excellent crystal quality, with low densities of dislocations, grain boundaries and V_O and Zn_i defects.

CONCLUSIONS

In conclusion, the effects of N_2O gas addition on the properties of zinc oxide (ZnO) films grown by a CVD method using high-temperature H_2O generated by a catalytic H_2-O_2 reaction on Pt-nanoparticles were investigated. Both the crystallinity and crystal orientations of films were improved by the addition of N_2O gas. The FWHM of the ZnO(0002) ω-rocking curve was decreased from 185 to 147 arcsec, while the electron mobility at room temperature was increased from 207 to 234 cm^2/Vs by the application of N_2O gas. The electron concentration of the ZnO film grown with N_2O decreased greatly at low temperatures and the activation energy calculated from variations in the electron concentration achieved an elevated value of 23.7 meV. PL spectra at room temperature showed a strong NBE band and a large NBE/DLE ratio in the case of the ZnO film grown with N_2O addition. Based on these results, the dislocations, grain boundaries and intrinsic donor impurities caused by defects are believed to be reduced by the addition of N_2O gas.

ACKNOWLEDGMENT

This work was supported in part by a Grant-in-Aid for Scientific Research (No. 24360014) from the Japan Society for the Promotion of Science.

REFERENCES

1. F. S. Hickernell, *Proc. IEEE*, **64**, 631 (1976).
2. S. Pizzini, N. Butta, D. Narducci, and M. Palladino, *J. Electrochem. Soc.*, **136**, 1945 (1989).
3. I. S. Jeong, J. H. Kim, and S. Im, *Appl. Phys. Lett.*, **83**, 2946 (2003).
4. T. Minami, *Semicond. Sci. Technol.*, **20**, S35 (2005).
5. B. K. Meyer, H. Alves, D. M. Hofmann, W. Kriegseis, D. Forster, F. Bertram, J. Christein, A. Hoffmann, M. Straßburg, M. Dworzak, U. Haboeck, and A. V. Rodina, *Phys. Stat. Sol. (b)*, **241**, 231 (2004).
6. P. Fons, K. Iwata, S. Niki, A. Yamada, and K. Matsubara: *J. Cryst. Growth*, **201–202**, 627 (1999).
7. K. Miyamoto, M. Sano, H. Kato, and T. Yao, *J. Cryst. Growth*, **265**, 34 (2004).
8. T. Ohgaki, N. Ohashi, H. Kanemoto, S. Wada, Y.Adachi, H. Haneda, and T. Tsurumi, *J. Appl. Phys.*, **93**, 1961 (2003).
9. M. Wei, R. C. Boutwell, N. Faleev, A. Osinsky, and W. V. Schoenfeld, *J. Vac. Sci. Technol. B*, **31**, 041206 (2013).

10. E. M. Kaidashev, M. Lorenz, H. von Wenckstern, A. Rahm, H. C. Semmelhack, K.-H. Han, G. Benndorf, C. Bundesmann, H. Hochmuth, and M. Grundmann, *Appl. Phys. Lett.*, **82**, 3901 (2003).
11. A. Ohtomo and A. Tsukazaki, *Semicond. Sci. Technol.*, **20**, S1 (2005).
12. J-L. Zhao, X-M. Li, J-M. Bian, W-D. Yu, and X-D. Gao, *J. Crystal Growth*, **276**, 507 (2005).
13. A. Tsukazaki, A. Ohtomo, T. Onuma, M. Ohtani, T. Makino, M. Sumiya, K. Ohtani, S. Chichibu, S. Fuke, Y. Segawa, H. Ohno, H. Koinuma, and M. Kawasaki, *Nature Materials*, **4**, 42 (2005).
14. R. M. Mundle, H. S. Terry, K. Santiago, D. Shaw, M. Bahoura, A. K. Pradhan, K. Dasari, and R. Palai, *J. Vac. Sci. Technol.* A, **31**, 01A146 (2013).
15. E. Fujimoto, M. Sumiya, T. Ohnishi, K. Watanabe, M. Lippmaa, Y. Matsumoto, and H. Koinuma, *Appl. Phys. Express*, **2**, 045502 (2009).
16. J. Dai, F. Jiang, Y. Pu, L. Wang, W. Fang, and F. Li, *Appl. Phys.*, A **89**, 645 (2007).
17. K. T. Roro, G. H. Kassier, J. K. Dangbegnon, S. Sivaraya, J. E. Westraadt, J. H. Neethling, A. W. R. Leitch, and J. R. Botha, *Semicond. Sci. Technol.*, **23**, 055021 (2008).
18. S. T. Tan, B. J. Chen, X. W. Sun, X. Hu, X. H. Zhang, and S. J. Chua, *J. Cryst. Growth*, **281**, 571 (2005).
19. K. Yasui, H. Miura, and H. Nishiyama, *MRS Symp. Proc.*, **1315**, 21 (2011).
20. J. Dai, H. Su, L. Wang, Y. Pu, W. Fang, and F. Jiang, *J. Cryst. Growth*, **290**, 426 (2006).
21. Y. Yan and S-H. Wei, *phy. stat. sol.* (b), **245**, 641 (2008).
22. Ü. Özgür, Y. I. Alivov, C. Liu, A. Teke, M. A. Reshchikov, S. Doğan, V. Avrutin, S.-J. Cho, and H. Morkoc, *J. Appl. Phys.*, **98**, 041301 (2005).

Density Functional Study of Benzoic Acid Derivatives Modified SnO_2 (110) Surface

Tegshjargal Khishigjargal and Kazuyoshi Ueda[*]
Department of Chemistry, Graduate School of Engineering, Yokohama National University 79-5 Tokiwadai, Hodogaya-ku, Yokohama, 240-8501, Japan

ABSTRACT

Tin oxide is one of the popular metal oxide semiconductor used in solar cells, sensors, and catalysts. The surface modification by organic self assembled monolayer is one of the promising techniques to tune and to control the surface work function. In our study, we investigated the work function change of the SnO_2 (110) surface which was modified with various benzoic acids derivatives using density functional theory (DFT). All calculations were carried out on Quantum Espresso program. Electron correlation and exchange parts were treated by local density (LDA), generalized gradient approximation (GGA) with Hubbard U term. To improve band structure calculation we used LDA+U method. The results of the calculation with LDA method indicated that the work functions of the pure and modified surface of SnO_2 (110) with -C_6H_4-COOH molecule were calculated to be 7.40 eV and 6.18 eV, respectively. As the experimental value of work function of SnO_2 (110) surface is about 7.74 eV, the results of the DFT calculation for pure SnO_2 (110) surface modification by benzoic acid derivatives are in good agreement with the experimental.

INTRODUCTION

Density functional theory (DFT) studies on metal oxide surfaces have recently been used to investigate the surface electronic structures and the reaction processes [1-9]. One of the widely investigated metal oxides is tin oxide (SnO_x). That is used for various fields, such as, gas sensors, solar cells, and catalysts and so on [10]. The surface of (110) rutile tin dioxide (SnO_2) is the most stable [11] and was studied by various models and DFT methods [1-6,8-10,12-15]. In the application of electronic devices, tin dioxide can be modified by the treatment of the self assembled monolayer (SAM) [16,17] and the work function of the surface can be tuned by this modification. Many experiments were carried out [18-21] to understand the nature of the WF change. However, the detail mechanism of the effect of the modification of the surface by the organic molecules has not been fully understood yet. Recently, the progress of the quantum chemical calculation enables us to understand the phenomena from the electronic structure point of view. In our previous work [22], work function change was investigated by quantum mechanical calculation using a cluster model of SnO_2. However, the estimated work function changes were higher than that of experimental measurements on benzoic acid SAM on indium-tin oxide (ITO) [17]. In this work, we investigated the WF change using the periodic DFT calculation with the model of benzoic acids (C_6H_4-COOH) on tin dioxide (110) surface. DFT calculation was performed with local density approximation (LDA) and generalized gradient exchange-correlation (GGA) functional.

CALCULATION METHOD

Tin dioxide with rutile structure was used as a crystal model in this study. The unit cell of SnO_2 consists of two tin and four oxygen atoms which has cell lengths of a=4.737 Å and c=3.186 Å [10]. The oxygen atoms are located in (±ua, ±ua, 0) for nearest tin atom with internal coordinate of u=0.307 [1] (Figure1. a). The model used in our study with 9-layers of the stochiometric SnO_2 (110) surface and the model with benzoic acid bound to the surface were shown in Figure 1. b and c, repectively. The orientation of the benzoic acids on SnO_2 (110) surface was the same as our previous model [22]. All atoms of the model were set free in relaxation. The relaxation criteria of atomic force is 0.001 a.u. The model with surface was cut out from the bulk crystal structure of SnO_2 after the relaxation of the conformation. All calculations were carried out by DFT methods with plane-wave basis set and ultra soft pseudopotential. To choose pseudopotentials, we performed some test calculations using ultrasoft (US) [23,24] and projector augmented wave (PAW) [25] for bulk SnO_2 optimization. For these pseudopotentials, the combination of the exchange correlation functional, Perdew-Zunger (PZ [26]), PZ+U (LDA+Hubbard U term [27-30]), Perdew-Burke-Ernzerhof (PBE [31]) and Perdew-Wang (PW91 [32]) within Quantum Espresso program were considered [33]. For Sn atom, 4d orbital was included in pseudopotential and Hubbard U term applied for oxygen atom with value of 5.3 eV. In the calculation of the inclusion of Hubbard U term, we tested the effect of the inclusion of U term to Sn atom. The results showed that there was no significant difference in the results of the optimized structures with and without Hubbard U term on Sn. WF was calculated using dipole correction [34] after the relaxed structure of benzoic acid on SnO_2 (110) slab. The k-point sampling grids of 4x4x4 and 4x4x1, obtained using Monkhorst-Pack method [35], are used for the SnO_2 bulk and SnO_2 (110) surface calculations, respectively. In the surface calculations all atoms were set free and compared other calculation results [12, 36, 37]. The kinetic energy cutoff of plane waves was 50 Ry and the charge density cutoff values for PAW and US pseudopotentials were 200 Ry and 400 Ry, respectively.

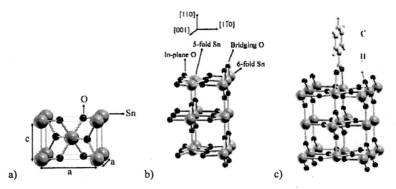

Figure 1. The models of Bulk SnO_2 (a), SnO_2 with (110) surface (b) and SnO_2 (110) surface with benzoic acid (c).

RESULTS AND DISCUSSION

Calculation of the bulk structure of SnO_2

The results of the structure optimization and the value of the band gap of the bulk SnO_2 calculated by various DFT methods were listed in Table I. The unit cell size of a and c are all within the deviation of 2 % from the experimental value. However, the calculated band gaps are all underestimated except the addition of Hubbard U term to the US_PZ method [27].

Table I. The optimized crystal structure of SnO_2 and band gap values.

	US_PBE	US_PW91	US_PZ	PAW_PBE	PAW_PZ	US_PZ+U	Exp. [11]
a (Å)	4.839	4.822	4.743	4.827	4.736	4.643	4.737
c (Å)	3.243	3.231	3.215	3.224	3.179	3.139	3.186
u	0.305	0.306	0.306	0.305	0.306	0.305	0.307
Dev. (a, c) (%)	2.15, 1.79	1.79, 1.41	0.13, 0.91	1.90, 1.19	0.02, 0.22	1.98, 1.48	
Band gap (eV)	0.55	0.72	0.94	0.65	1.07	3.59	3.6

As an example, the band structures calculated with US_PZ and US_PZ+U methods are compared in Figure 2. It shows that the band gap at Γ point calculated with US_PZ method is 0.98 eV. As where shown in Table I, this value is close to those obtained from the calculation with other DFT methods [8]. On the other hand, the band gap increased linearly when we added U to the US_PZ method. The increment is almost linearly proportional to the value of U term. Instead, the unit cell size shrank and deviation became large when U value increased. As a best value of U, we selected the value of U=5.3 eV which gave the band gap value of 3.59 eV at Γ point. It should be mentioned that the calculations with PAW_PZ showed the least deviation of the unit cell size from the experimental values. However, further apply of this method to the model of SnO_2 with (110) surface could not be converged in our calculation. Therefore, we used US_PZ and US_PZ+U methods for our further calculations on the model of SnO_2 crystal with (110) surface for the optimization and the band gap calculation, respectively.

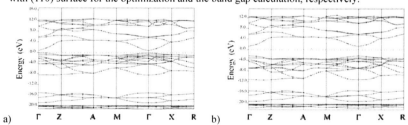

Figure 2. The whole band structures of bulk SnO_2 calculated by US_PZ (a) and US_PZ +U (b).

Calculation of the SnO_2 crystal with (110) surface and surface with benzoic acid modification

After the optimization of the bulk crystal structure with US_PZ method, 9-layers of stochiometric SnO_2 were cut out and the crystal with (110) surface was built as was shown in Figure 1. b. In this model, unit cell size were chosen as 2c (6.430 Å), a√2 (6.708 Å), and 40 Å along to the [001], [1$\bar{1}$0], and [110] directions, respectively. The results of the structure optimization with US_PZ method was shown in Table II. In this table, the reported results calculated using with other methods are also shown as a comparison. The atoms of first two layers were found to be displaced toward the vacuum region (Table 2).

Table II. Displacement (Å) along [110] direction of SnO_2 (110) surface atoms (1st and 2nd layers). The displacements were measured as relative distance from 5-fold Sn atom on second layer after the relaxation by US_PZ method.

	In-plane O (2nd layer)	Bridging O (1st layer)	6-fold Sn (2nd layer)	5-fold Sn (2nd layer)
US_PZ	0.24	0.07	0.27	0.00
US_PBE [12]	0.18	0.09	0.22	-0.11
TB [36]	0.29	0.06	0.10	-0.12
PWPP [37]	0.22	0.17		0.30

Although the displacements of the atoms in Table II were shown from the 5-fold Sn atom, 5-fold Sn atom also moved toward the vacuum region by 0.32 Å from the bulk positions. We need to further investigate the calculation methods to well reproduce the SnO_2 (110) surface.

The structure optimization of the surface model with benzoic acid SAM layer (Figure 1. c) was performed. The benzoic ring plane showed perpendicular orientation toward the surface and no tilting from the [110] direction of the surface. However, the benzoic ring plane rotated 28.2° from the original *cis* angle with carboxyl oxygen (-COO) plane which bound to the surface of SnO_2 [22].

To estimate work function change, which was occurred by the modification with benzoic acid on the surface, plane averaged electrostatic potential (Figure 3) was calculated from relaxed geometry of benzoic acid on SnO_2 (110) slab. In this calculation, dipole correction [34] was applied to cancel the effect which was caused by the asymmetry of charges on two sides of the calculation model in Figure 1. c.

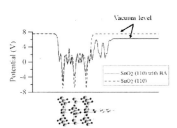

Figure 3. Plane averaged electrostatic potential of SnO_2 (110) surface with (solid line) and without benzoic acid (dashed line). Fermi energy levels are set as zero.

In Figure 3, potential curve of 9-layers SnO_2 (110) surface is shown in dashed line, which provides the value of WF of 7.40 eV from the vacuum and Fermi level difference. This value is in good accordance with the experimental value of 7.74 eV [36]. After the modification of the surface by benzoic acid, vacuum level decreased and work function value was calculated to be 6.18 eV. From these values, the work function change of SnO_2 (110) surface generated by the modification with benzoic acid was estimated to be 1.32 eV. As a summary, it was shown in this work that the work function for oxide metal surface can well be evaluated by using DFT methods. Further detailed calculations of WF using various kinds of benzoic acid derivatives will be discussed in the future.

CONCLUSIONS

DFT computational calculation using models of bulk SnO_2, slab SnO_2 with (110) surface, and slab SnO_2 with (110) surface modified by SAM treatment of benzoic acid were reported in this work. For the calculation of work function with periodic DFT method in tin dioxide (110) surface showed in good agreement with the experimental value. We also succeeded to evaluate the work function change of SnO_2 with SAM by the treatment of benzoic acid. However, as far as we know, there is no experimental data of the WF change which measured the SAM treatment SnO_2 surface. If we use the experimental results of Ganzorig et al., [18] who measured the benzoic acid modification on ITO, work function decreased of 0.2 (±0.05) eV from the pure ITO surface. In our calculation, work function change showed higher value compared to this experimental value, but showed the same tendency on ITO surface modification by benzoic acid. For the quantitative evaluation of our calculation, future experimental work will be desired. From the point of the structure relaxation, SnO_2 (110) surface atoms displaced a little from the bulk structure, we need to seek more accurate calculation methods which we will need to consider to use other functional, such as, PBE and PW91 [2-9].

ACKNOWLEDGMENTS

The computations were partly performed using Research Center for Computational Science, Okazaki, Japan.

REFERENCES

1. S. Munnix and M. Schmeits, Phys. Rev. B **27**, 7624 (1983).
2. I. Manassidis, J. Goniakowski, L.N. Kantorovich, M.J. Gillan, Surf. Sci. **339**, 258 (1995).
3. M.A. Mäki-Jaskari, T.T. Rantala, Phys. Rev. B **65**, 245428 (2002).
4. T.T. Rantala, T.S. Rantala, V. Lantto, Surf. Sci. **420**, 103 (1999).
5. J. Oviedo, M.J. Gillan, Surf. Sci. **463**, 93 (2000).
6. M. Ramamoorthy, D. Vanderbilt, R.D. King-Smith, Phys. Rev. B **49**, 16721 (1994).
7. S.P. Bates, M.J. Gillan, G. Kresse, J. Phys. Chem. B **102**, 2017 (1998)
8. M. Calatayud, J. Andrés, and A. Beltrán, Surf. Sci. **430**, 213 (1999).
9. Y. Yamaguchi, K. Tabata, T. Yashima, J. Mol. Str: THEOCHEM **714**, 221 (2005).
10. M. Batzil and U. Diebold, Prog. Surf. Sci. **79**, 47 (2005).
11. V.E. Henrich, P.A. Cox, The Surface Science of Metal Oxides, Cambridge University Press, Cambridge, (1994).
12. M. Viitala, O. Cramariuc, T. T. Rantala, and V. Golovanov, Surf. Sci. **602**, 3038 (2008).

13. Z. Wen, L. Tian-mo, and L. Xiao-fei, Physica B **405**, 3458 (2010).
14. M. Melle-Franco and G. Pacchioni, Surf. Sci. **461**, 54 (2000).
15. Y. Duan, Phys. Rev. B **77**, 045332 (2008).
16. M. Fujihara, N. Ohishi, and T. Osa, Nature **268**, 226.
17. C. Ganzorig and M. Fujihara, "Chemically Modified Oxide Electrodes. Encyclopedia of Electrochemistry," (Wiley-VCH Verlag GmbH & Co. KGaA, 2007).
18. Ch. Ganzorig, K.-J. Kwak, K. Yagi, and M. Fujihira, Appl. Phys. Lett. **79**, 272 (2001).
19. M. Carrara, F. Nüesch and L. Zuppiroli, Synth. Met. **121**, 1633 (2001).
20. F. Nüesch, F. Rotzinger, L. Si-Ahmed and L. Zuppiroli, Chem. Phys. Lett. **288**, 861 (1998).
21. S. F. J. Appleyard, S. R. Day, R. D. Pickford and M. R. Willis, J. Mater. Chem. **10**, 169 (2000).
22. T. Khishigjargal, N. Javkhlantugs, C. Ganzorig, Y. Kurihara, M. Sakomura, K. Ueda, World Journal of Nano Science and Engineering **3**, 52-56 (2013).
23. G. Kresse and D. Joubert, Phys. Rev. B **59**, 1758 (1999).
24. D. Vanderbilt, Phys. Rev. B **41**, 7892 (1990).
25. P.E. Blochl, Phys. Rev. B **50**, 17953 (1994).
26. J.P. Perdew and A. Zunger. Phys. Rev. B **23**, 5048 (1981).
27. V. I. Anisimov, J. Zaanen, and O.K. Andersen, Phys. Rev. B **44**, 943 (1991).
28. V.I. Anisimov, I.V. Solovyev, M.A. Korotin, M.T. Czyżyk and G.A. Sawatzky, Phys. Rev. B **48**, 16929 (1993).
29. A.I. Liechtenstein, V.I. Anisimov, and J. Zaanen, Phys. Rev. B. **52**, R5467 (1994).
30. M. Cococcioni and S. de Gironcoli, Phys. Rev. B **71**, 035105 (2005).
31. J.P. Perdew, K. Burke, and M. Ernzerhof, Phys. Rev. Lett. **77**, 3865 (1996).
32. J.P. Perdew, J.A. Chevary, S.H. Vosko, K.A. Jackson, M.R. Pederson, D.J. Singh, and C. Fiolhais, Phys. Rev. B **46**, 6671 (1992).
33. P. Giannozzi, S. Baroni, N. Bonini, M. Calandra, R. Car, C. Cavazzoni, D. Ceresoli, G.L. Chiarotti, M. Cococcioni, I. Dabo, A. Dal Corso, Stefano de Gironcoli, S. Fabris, G. Fratesi, R. Gebauer, U. Gerstmann, C. Gougoussis, A. Kokalj, M. Lazzeri, L. Martin-Samos, N. Marzari, F. Mauri, R. Mazzarello, S. Paolini, A. Pasquarello, L. Paulatto, C. Sbraccia, S. Scandolo, G. Sclauzero, A. P. Seitsonen, A. Smogunov, P. Umari, and R. M. Wentzcovitch, J. Phys.: Conden. Matt. **21**, 395502 (2009).
34. L. Bengtsson, Phys. Rev. B **59**, 12301 (1999).
35. H.J. Monkhorst and J.D. Pack, Phys. Rev. B **13**, 5188 (1976).
36. T.J. Godin and J.P. LaFemina, Phys. Rev. B **47**, 6518 (1993)..
37. J. Goniakowski, J. M. Holender, L. N. Kantorovich, and M. J. Gillan, Phys. Rev. B **53**, 957 (1996).

Defect Driven Emission from ZnO Nano Rods Synthesized by Fast Microwave Irradiation Method for Optoelectronic Applications

Nagendra Pratap Singh[1,2], S.A. Shivashankar[2], Rudra Pratap[1,2]

[1]Department of Mechanical Engineering, Indian Institute of Science, Bangalore-560012, India
[2]Centre for Nano Science and Engineering (CeNSE), Indian Institute of Science, Bangalore-560012, India

ABSTRACT

Because of its large direct band gap of 3.37 eV and high exciton binding energy (~60 meV), which can lead to efficient excitonic emission at room temperature and above, ZnO nanostructures in the würtzite polymorph are an ideal choice for electronic and optoelectronic applications. Some of the important parameters in this regard are free carrier concentration, doping compensation, minority carrier lifetime, and luminescence efficiency, which are directly or indirectly related to the defects that, in turn, depend on the method of synthesis. We report the synthesis of undoped ZnO nanorods through microwave irradiation of an aqueous solution of zinc acetate dehydrate [$Zn(CH_3COO)_2 \cdot 2H_2O$] and KOH, with zinc acetate dihydrate acting as both the precursor to ZnO and as a self-capping agent. Upon exposure of the solution to microwaves in a domestic oven, ZnO nanorods 1.5 μm -3 μm and 80 nm in diameter are formed in minutes. The ZnO structures have been characterised in detail by X-ray diffraction (XRD), selective area electron diffraction (SAED) and high-resolution scanning and transmission microscopy, which reveal that each nanorod is single-crystalline. Optical characteristics of the nanorods were investigated through photoluminescence (PL) and cathodoluminescence (CL). These measurements reveal that defect state-induced emission is prominent, with a broad greenish yellow emission. CL measurements made on a number of individual nanorods at different accelerating voltages for the electrons show CL intensity increases with increasing accelerating voltage. A red shift is observed in the CL spectra as the accelerating voltage is raised, implying that emission due to oxygen vacancies dominates under these conditions and that interstitial sites can be controlled with the accelerating voltage of the electron beam. Time-resolved fluorescence (TRFL) measurements yield a life time (τ) of 9.9 picoseconds, indicating that ZnO nanorods synthesized by the present process are excellent candidates for optoelectronic devices.

INTRODUCTION

Zinc oxide (ZnO) is a multifunctional, direct, wide band gap (E_g=3.37 eV) semiconductor with a large free-exciton binding energy (~60 meV), rendering excitonic emission processes possible at or above room temperature[1,2,3]. Thus, ZnO is a promising material for UV and blue light emitting devices with several fundamental advantages over GaN, its chief competitor. It is more radiation-resistant than GaN, comparatively inexpensive, and biocompatible [4,5]. Defects in ZnO play a major role in tailoring its optical, piezoelectric, mechanical, electronic and optoelectronic properties [3]. The desired properties of ZnO nanostructures can be tuned with either native defects or foreign atom incorporated defects (doping). Intrinsic point defects in ZnO nanorods give rise to luminescence over a broad range in the visible spectrum. This enables tunable resonant emission in the visible spectral range with a pure ZnO system. Native defect engineering is favorable over doping because foreign atoms usually affect surface morphology as well as physical, chemical, and optical properties to a greater degree than do native defects. As these are point defects created generally by either oxygen or zinc vacancies, their formation depends on the process of synthesis [6]. Among solution-based processes, microwave-assisted synthesis (MWS) is relatively simple and rapid, and allows control over the growth process through numerous parameters, including the choice of precursor and solvent, and can yield device quality nanomaterial inexpensively compared to other growth processes such as CVD, VLS, PVD, and hydrothermal [7]. We have used simple, surfactant-free chemistry to synthesize high quality ZnO nanorods in the aqueous medium. The tuning of the optical properties is demonstrated with the help of cathodoluminescence (CL) spectra obtained at varying accelerating voltages of incident electrons. The possibility of optical devices requiring fast switching is indicated as a carrier lifetime in these nanorods is found to be less than 10 ps.

EXPERIMENT

To synthesize ZnO nanorods, zinc acetate dihydrate and KOH (in 1:15 molar ratio) is dissolved in 100 ml of de-ionised water (18 MΩ-cm). The solution is stirred for 15 min in a round bottom flask and exposed to microwave radiation at 800 Watt power for 3 minutes in a domestic microwave oven (LG MS 2049, 2.45 GHz). The resulting white precipitate is separated by centrifugation at 4000 rpm and washed in ultra-pure water. The precipitate is then dried in an oven at 75^{0}C. The resulting powder is analyzed using X-ray powder diffraction (XRD), field-emission scanning electron microscopy (FESEM), transmission electron microscopy (TEM), photoluminescence (PL, Horiba LabRAM HR), and cathodoluminescence (CL, FESEM-Carl Zeiss Mono). Time-resolved fluorescence lifetime measurement (TRFL, Horiba JobinYvon Fluorocube-01-NL Fluorescence Life time System) was made to determine carrier lifetime at room temperature. A picosecond laser diode (374 nm wavelength) was used as the excitation source.

RESULTS and DISCUSSION

The XRD pattern recorded with a Rigaku powder diffractometer is shown in Fig. 1 (a). All the diffraction peaks can be indexed to zinc oxide of the hexagonal würtzite structure. The XRD data yield lattice constants of a=3.2478 Å and c =5.2035 Å [JCPDS –36-1451]. The Scherrer formula [8] applied to the (101) peak gives the crystallite size of ZnO to be 62.5 nm, equivalent to the diameter of the nanorods. The average crystallite size (D= 62.5 nm) was calculated using the Scherrer formula,

$$D = \frac{0.9\lambda}{\beta_{hkl}\cos\theta} \quad \ldots \ldots \ldots \ldots \ldots \ldots \ldots \ldots \ldots \ldots \ldots \ldots \ldots \ldots \ldots \ldots \ldots (1)$$

where β_{hkl} is the full width at half maximum (FWHM) and λ=1.54 Å.

The Williamson-Hall equation [9] is used to estimate the strain in the ZnO crystallites:

$$\beta_{hkl}\cos\theta = \frac{K\lambda}{D} + 4\,\varepsilon\,\sin\theta \quad \ldots \ldots \ldots \ldots \ldots \ldots \ldots \ldots \ldots \ldots \ldots \ldots \ldots \ldots (2)$$

where ε the lattice strain and K (=0.9) is a constant. $\beta\cos\theta$ is plotted against $4\sin\theta$ in Fig. 1 (b). A linear fit to the data yields the crystallite size D through the intercept $\frac{K\lambda}{D}$ and the strain (ε) is given by the slope of the straight linefrom the graph (Fig. 1b). The microstrain ε = 0.097%, which is indicative of the strain due to native defects present in the crystals.

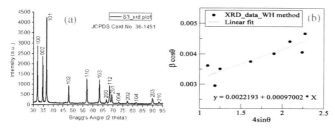

Figure 1 (a) Powder X-ray diffraction pattern of ZnO nanorods, all peaks corresponding to hexagonal ZnO; **(b)** Williamson-Hall plot derived from the XRD data

Figure 2 FESEM image of ZnO nanorods

FESEM analysis (Fig.2) reveals clearly that the ZnO sample is made of hexagonal nanorods. It is seen that the rods are well-faceted, but are "tapered" at one end, indicative of the rather gradual termination of the growth process upon microwave power being turned off. They range in length from ~1.5 μm to 3 μm and in diameter from ~30 nm to 80 nm. Most rods have relatively large diameters, consistent with the average crystallite size estimated from XRD data. TEM analysis (Fig. 3 (a), 3 (b)), including high-resolution imaging and selected-area electron diffraction (SAED) together reveal that each nanorod is a single crystal of high quality and purity. HRTEM and SAED show also that the ZnO nanorods are oriented in the <0001> direction.

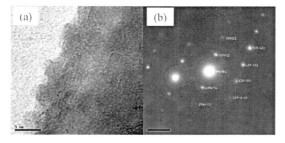

Figure 3 (a) HRTEM image (b) SAED pattern of a ZnO nanorod

The photoluminescence spectrum (Fig. 4a) obtained from an assembly of ZnO nanorods at room temperature through excitation with a He-Cd laser (325 nm) shows that the near-band-edge emission at ~3.27 eV is dwarfed by the strong broadband emission in the green, centred at ~2.20 eV. The components of this well-known emission in ZnO due to native defects are revealed by deconvolution (Fig.4 (b)). As the schematic band diagram (Fig.4 (c)) shows, the components of the broadband emission are all from deep within the bandgap in the undoped ZnO sample, the Fermi energy being ~1.64 eV. Emission in the green from ZnO is known to be due to oxygen vacancies [10].

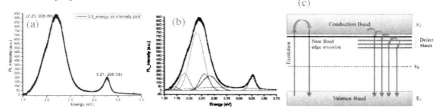

Figure 4(a) Photoluminescence spectrum, (b) Photoluminescence spectrum with de-convoluted peaks and (c) Schematic model for different relaxation processes in ZnO nanostructures

Cathodoluminescence (CL) allows control over emission from native defects through control over such defects provided by the acceleration potential of the electron beam in the SEM. The CL spectra of the ZnO nanostructures obtained with the electron beam accelerated to different voltages are shown in Figs.5 (a) and 5(b). The sample was anchored carefully to a substrate to eliminate artifacts due to heating by the e-beam. A clear – though small - "red shift" of 2.5 nm (between 5 kV and 20 kV of the acceleration potential) in the emission due to the defects as the electron beam is accelerated through higher potentials can be discerned from these spectra. A monotonic increase in the intensity of the broadband emission is seen as the accelerating potential in increased, accompanied by a corresponding monotonic reduction in the intensity of the near-band-edge emission. These changes in intensity together provide evidence that the density of native defects (related to oxygen in ZnO) increases, with an inevitable deterioration in the degree of perfection in the crystals, on which the intensity of the near-band-edge emission depends.

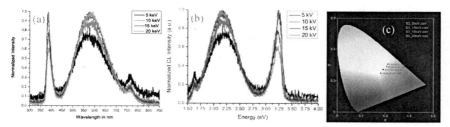

Figure 5(a,b) Cathodoluminescence (CL) spectrum at varying voltages; red shift in the defect peak shows that emission characteristics are tunable with voltage; sample carefully anchored to substrate to avoid heating due to e-beam. (c) CIE color map of CL spectrum.

In a different manifestation of the changes in the ZnO defect structure under electron bombardment, the colour of cathodoluminescence changes steadily as the acceleration potential is increased. This is shown in the CIE colour map (Fig. 5 (c)), with the emission turning from green to orange, with a near-white emission at the acceleration potential of 10 kV.

To examine the process of luminescence from ZnO further, the lifetime of photo-generated carriers was measured by the time-resolved fluorescence technique, with the sample excited by a laser diode emitting at 374 nm, with the lifetime measured at 575 nm.

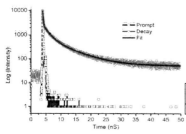

Table 1: Luminescence decay parameters obtained with an excitation wavelength at 374 nm, measured at 575 nm and τ the lifetime and A & B the corresponding relative amplitude

A	B_1	τ_1 (ps)	B_2	τ_2 (ps)	B_3	τ_3 (ps)	χ^2
51.4	33.6	1846.4	49.1	7260.5	17.2	191.1	1.8

Figure 6 Time-resolved fluorescence life time of ZnO nanorods excited by a 374 nm laser diode and measured at 575 nm. Carrier lifetime is calculated to be 191 picoseconds

The intensity of the transient emission (Fig.6) as a function of time (in picoseconds) was fitted to a sum of exponentials [11], according to the equation (1):

$$I(t) = \sum_{i=1}^{3} \alpha_i e^{-\frac{t}{\tau_i}} \quad\quad\quad\quad\quad\quad\quad\quad\quad\quad\quad\quad\quad\quad\quad\quad\quad\quad\quad\quad (3)$$

where i is the number of exponentials, and τ_i and α_i are the lifetime and the pre-exponential factor for each component. The TRFL decay spectra of the ZnO nanorods with exponential fits are given in Figure 6 and fitted parameters are listed in table-1. After finding the best fit, we deduce the lifetime of the carriers in the ZnO nanorod to be 191 picoseconds. This indicates fast switching, which shows the potential of this material for applications in memory devices, LEDs, and laser diodes.

CONCLUSIONS

We have synthesized ZnO nanorods by the rapid and inexpensive microwave irradiation method. Structural, morphological, and optical characterization show them to be single-crystalline and of high quality. The tunability of their properties through engineering the defect structures and defect density has been demonstrated by subjecting them to irradiation by electron beams accelerated to different potentials. For example, the colour of emission (in the visible) due to oxygen defects is found to be controllable by just tuning the excitation voltage. Further, carrier lifetime in these nanostructures is found to be in the sub-nanosecond range, which indicates rapid generation and recombination of carriers. Thus, a voltage-controlled, single nanorod-based optoelectronic device appears to be realizable.

ACKNOWLEDGMENTS

This work has been supported by CEN Phase II grant from the Department of Electronics and Information Technology, Government of India. Partial support provided by Centum Electronics, a member of the Industry Affiliate Program, enabling NPS to attend the MRS Fall Meeting, is gratefully acknowledged.

REFERENCES

[1] A. H. Jayatissa, A. M. Soleimanpour, and Y. Hao, *Adv. Mater. Res.*, **383–390**, pp. 4073–4078, (2011).

[2] A. Janotti and C. G. V. de Walle., *Reports Prog. Phys.*, **72**, pp. 126501-126530, (2009).

[3] H. M. and U. Ozgur, *Zinc Oxide: Fundamentals, Materials and Device Technology*, Wiley-VCH Verlag GmbH & Co. 2007, pp. 490.

[4] Z. L. Wang and J. Song, *Science*, **312**, pp. 242–6, (2006).

[5] Z. Li, R. Yang, M. Yu, F. Bai, C. Li, Z. L. Wang , *J. Phys. Chem. C*, **112**, pp. 20114–20117, (2008).

[6] D. F. Wang and T. J. Zhang, *Solid State Commun.*, **149**, pp. 1947–1949, (2009).

[7] S. Brahma, K. J. Rao, and S.A. Shivashankar, *Bull. Mater. Sci.*, **33**, pp. 89–95, (2010).

[8] B. D. Cullity, *The Elements of X-Ray Diffraction*, Addison-Wesley, 1978, pp. 102.

[9] Y. P. and B. D. V.D. Mote, *J. Theor. Appl. Phys.*, **6**, pp. 2–9, (2012).

[10] A. B. Djurisic and Y. H. Leung, *Small*, **2**, pp. 944-961, (2006).

[11] E. Sönmez and K. Meral, *J. Nanomater.*, **2012**, pp. 1–6, (2012).

Breaking of Raman selection rules in Cu_2O by intrinsic point defects

Thomas Sander[1,*], Christian T. Reindl[1], and Peter J. Klar[1]

[1] I. Physikalisches Institut, Justus-Liebig-Universität Gießen, Heinrich-Buff-Ring 16, 35392 Gießen, Germany

*Thomas.Sander@physik.uni-giessen.de

ABSTRACT

The semiconductor cuprous oxide crystallizes in a simple cubic structure and reveals outstanding characteristics: Independent of the method and conditions of the synthesis of crystalline Cu_2O its Raman spectra are dominated by infrared active, silent, and defect modes rather than by Raman allowed phonon modes only. A detailed group theoretical analysis demonstrates that point defects reduce the local symmetry, lift the Raman selection rules, and thus diminish the distinction between Raman allowed and Raman forbidden lattice vibrations. Of all intrinsic defects only the presence of the copper vacancy in the so called split configuration introduces possible Raman activity for all Cu_2O extended phonon modes observed in experiment.

INTRODUCTION

Already in 1951, in a review article about copper oxide rectifiers, W. H. Brattain has described the origin of Cu_2O as a semiconductor by "Copper oxide is a defect semiconductor. ... The main impurity centers, acceptors in this case, are probably vacant copper ion lattice sites." [1]. Hardly anything has changed concerning the validity of this statement, nowadays it is well established that Cu_2O is a natural p-type semiconductor, whose carrier concentration depends on the amount of cation deficiency (non-stoichiometry).

Non-stoichiometry due to the formation of point defects such as vacancies, interstitials, or antisite defects has three major effects in Raman spectroscopy: On the one hand, perfect translational symmetry is broken leading to the breakdown of the wave vector selection rules allowing phonon momentum values from the whole Brillouin zone. Further, the Raman selection rules do no longer hold strictly, e.g. the point defects reduce the local symmetry such that the distinction between Raman allowed and forbidden lattice vibrations diminishes compared to the space symmetry of the crystal. On the other hand, depending on the point defect and its compatibility with the lattice, local vibrational modes may be introduced, which may also be Raman active. The Raman spectra of Cu_2O are very good examples for these effects as the dominant Raman signals observed are actually due to infrared active, silent, or defect modes rather than due to the nominally Raman active mode.

VIBRATIONAL CHARACTERISTICS OF Cu_2O

Three different phases CuO (cupric oxide), Cu_2O (cuprous oxide), and Cu_4O_3 (paramelaconite) of the binary semiconductor are known. Each reveals unique Raman spectra due to their differences in crystal structure [2]. As depicted in figure 1a), cuprous oxide crystallizes in a simple cubic structure of space group O^4_h (Pn-3m) [3,4]. Its unit cell contains two Cu_2O units, i.e. six atoms, yielding 18 phonon modes. Huang performed the first group theoretical analysis of the vibrational modes already in 1963 [5]. The symmetries of the vibrational modes at the center of the Brillouin zone are given by:

$$A_{2u} \oplus E_u \oplus T_{2u} \oplus T_{2g} \oplus 3T_{1u} . \qquad (1)$$

Modes with E and T symmetry are two- and three-fold degenerate, respectively. The three acoustic phonons possess T_{1u} symmetry. Optical phonons with T_{1u} symmetry are infrared-active. The A_{2u} and E_u modes are silent modes, neither Raman nor infrared active. Vibrations with T_{2g} symmetry are the only Raman active modes in this structure.

However, typical Raman spectra of Cu_2O shown in figure 2a) are much richer with a multitude of Raman signals assigned to infrared active or silent modes labeled according to the irreducible representations of the corresponding phonon modes in O^4_h symmetry. The spectra shown are taken of various Cu_2O samples, i. e. a bulk crystalline sample as well as thin film samples grown by molecular beam epitaxy (MBE), chemical vapour deposition (CVD), rf-sputtering, and oxidized copper recorded at room temperature in backscattering geometry using a 532 or 633 nm laser for excitation. Very similar spectra of bulk Cu_2O crystals are reported in the literature [6-13]. The main finding basically is that one observes all lattice modes as one-phonon Raman signals in the spectra: a T_{2u} mode at about 90 cm^{-1}, a E_u mode at 110 cm^{-1}, LO and TO modes of T_{1u} symmetry between 140-160 cm^{-1}, a A_{2u} mode at about 350 cm^{-1}, and LO and TO modes of T_{1u} symmetry between 630-660 cm^{-1}, and in the vicinity of 515 cm^{-1} the only Raman active LO and TO T_{2g} modes. Additional features at 220 cm^{-1} and in the range between 400 and 490 cm^{-1} are assigned to a two-phonon process ($2E_u$) and multi-phonon Raman scattering, respectively.

GROUP THEORETICAL DISCUSSION

According to the group theoretical analysis above, a perfect Cu_2O crystal should exhibit only the three-fold degenerate Raman active T_{2g} mode. Due to non-stoichiometry the Raman selection rules may no longer strictly hold. In particular, the symmetry of the site occupied by a point defect defines a local symmetry which is usually lower than that of the space group of the crystal. As a consequence the selection rules for the perfect crystal, e. g. Raman selection rules for extended phonon states, do no longer hold strictly. Vibrational modes which are Raman forbidden in a perfect crystal may become Raman allowed due to the reduction of symmetry caused by the point defect. The symmetry of displacement pattern of the phonon eigenstate of the perfect crystal, which determines whether a mode is silent, Raman or infrared allowed, is represented by an irreducible representation of the space group of the crystal. Reducing the symmetry to the site symmetry of the point defect may make this representation reducible, i. e. it may be expanded in terms of the irreducible representations of the point group characterizing the

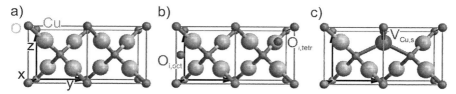

Figure 1 Schematic illustration of two primitive unit cells of a perfect Cu_2O crystal in a). Oxygen interstitial defects and the copper split vacancy are sketched in b) and c), respectively.

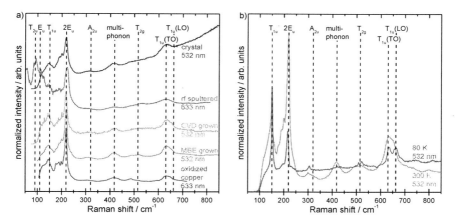

Figure 2 In a) Raman spectra at room temperature of various Cu_2O samples grown by different methods are shown. For clarity, the spectra are shifted on the y-axis. In b) Raman spectra of MBE grown Cu_2O at room temperature and 80 K are compared. The Raman signals are labeled according to the irreducible representations of the corresponding phonon modes in O_h symmetry.

defect. The irreducible representations of the point group of the defect site in the expansion of the representation of the original phonon state may exhibit different silent, Raman or infrared active behavior than that of the original mode. In case of degenerate modes the expansion may be a sum of different irreducible representations, e. g. formerly triply-degenerate T modes may be represented by a sum of an A mode and a doubly-degenerate E mode which may even have different frequencies.

A group theoretical analysis of this kind has been performed by Reydellet *et al.* in Ref. 6 for substitutional defects on copper and oxygen sites of Cu_2O. As depicted in table I, all phonon modes at the Γ-point with the exception of the former T_{2u} mode become Raman active for a substitutional defect on an oxygen lattice site Cu_O. However, in the experimental Raman spectra the T_{2u} phonon exhibits a prominent feature at about 90 cm^{-1} (see figure 2). To explain the observation of the T_{2u} phonon among others Reydellet *et al.* extended their analysis

Table I. Effect of reduced local symmetry in the vicinity of different point defects on the Raman activity of the extended phonon modes at the Γ-point of Cu_2O. The phonons of ideal Cu_2O are labeled according to the irreducible representations of the O_h group. The representations of O_h are expanded in terms of irreducible representations of the point group describing the symmetry of the defect site. Irreducible representations of Raman allowed modes are plotted in bold.

point defect				reduction of the vibrational modes at the Γ-point in O_h symmetry due to the defect site-symmetry				
symbol	Wyckoff position	multiplicity	site symmetry	A_{2u}	E_u	T_{2u}	T_{2g}	T_{1u}
V_O	a	2	T_d	$\mathbf{A_1}$	E	T_1	T_2	T_2
Cu_O	a	2	T_d	$\mathbf{A_1}$	E	T_1	T_2	T_2
V_{Cu}	b	4	D_{3d}	A_{1u}	E_u	$A_{1u} \oplus E_u$	$\mathbf{A_{1g} \oplus E_g}$	$A_{1u} \oplus E_u$
O_{Cu}	b	4	D_{3d}	A_{1u}	E_u	$A_{1u} \oplus E_u$	$\mathbf{A_{1g} \oplus E_g}$	$A_{1u} \oplus E_u$
$O_{i,oct}$	c	4	D_{3d}	A_{1u}	E_u	$A_{1u} \oplus E_u$	$\mathbf{A_{1g} \oplus E_g}$	$A_{1u} \oplus E_u$
$O_{i,tetr}$	d	6	D_{2d}	$\mathbf{A_1}$	$\mathbf{A_1 \oplus B_1}$	$A_2 \oplus E$	$\mathbf{B_2 \oplus E}$	$\mathbf{B_2 \oplus E}$
$V_{Cu,s}$	f	12	D_2	$\mathbf{A_1}$	$\mathbf{2 A_1}$	$\mathbf{B_1 \oplus B_2 \oplus B_3}$	$\mathbf{B_1 \oplus B_2 \oplus B_3}$	$\mathbf{B_1 \oplus B_2 \oplus B_3}$

to other critical points of the Brillouin zone which finally lifted the Raman forbiddance of the T_{2u} mode. This assumption requires that the wave vector selection rules are broken like in amorphous materials. Since any kind of bulk crystal possesses a certain amount of intrinsic point defects broadened and inactive Raman features should be observable in the Raman spectra of any crystal. This may be true to some extent, however, in case of Cu_2O the features remain fairly sharp somewhat contradicting the view of weakened k-selection rules.

In the following, we will extent the analysis to other recently predicted intrinsic point defects of Cu_2O, in particular, the two types of oxygen interstitials and the copper vacancy in split configuration $V_{Cu,s}$, schematically illustrated in figure 1b) and c), respectively. Theoretical studies dealing with intrinsic point defects in cuprous oxide argue that the copper vacancy and the copper split vacancy are most likely responsible for the natural p-type conductivity of Cu_2O as their formation energy is lower than that of other possible intrinsic acceptors [13-17]. Table I summarizes results of our group theoretical analysis of the effect of the different intrinsic point defects of Cu_2O in terms of defect-induced Raman activity of the extended phonon modes at the Γ-point (k=0) of the ideal crystal. The irreducible representations stand for the behavior of the corresponding phonon modes. Bold printed representations indicate Raman allowed modes. The irreducible representations given in the header of the table are those valid in O_h symmetry of the perfect crystal. The column below each original representation in O_h symmetry shows its expansions in terms of the irreducible representations of the point groups representing the site symmetry of the various intrinsic defects.

Point defects on copper site (Wyckoff position b), i.e. V_{Cu} and O_{Cu}, as well as the oxygen interstitial $O_{i,oct}$ (Wyckoff position c) possess D_{3d} site symmetry and have the smallest impact on the Raman spectra. The vibrational modes which are Raman forbidden in the ideal crystal remain

Raman forbidden, whereas the representation T_{2g} of the Raman active mode is expanded into the two Raman allowed irreducible representations A_{1g} and E_g of D_{3d}. The point defects V_O and Cu_O related to the oxygen sites of the unit cell (Wyckoff position a) lift the Raman forbiddance for all phonon modes at Γ with the exception of the former T_{2u} mode. Somewhat similar are the effects of the oxygen interstitial $O_{i,tetr}$ (Wyckoff position d). Again all modes with the exception of the T_{2u} mode become Raman active. In addition, mode splitting for the degenerate modes is introduced, however, in all cases but for the former E_u mode only one of the new irreducible representations is Raman allowed. Of all intrinsic defects discussed, only the presence of the $V_{Cu,s}$ point defect introduces possible Raman activity for all Cu_2O extended phonon modes at the Γ-point. In addition, all originally degenerate modes split, i. e. the original E_u mode into two A modes and each original T mode into a sum of B_1, B_2, and B_3 modes. As depicted in figure 2b), the T_{2g} and T_{1u} modes split into separate signals at low temperatures which might indicate that the three-fold degeneracy is lifted into different B modes or that LO and TO signals get separated.

CONCLUSIONS

Copper oxide takes an outstanding role in Raman spectroscopy since it is dominated by Raman forbidden phonons no matter which growth technique is applied. Reydellet et al. explained the experimental findings by defect activated modes of point defects on Cu and O sites from different critical points of the Brillouin zone [6]. This explanation requires a significant breaking of k-selection rules. Recent theoretical work predicted more complex defects in Cu_2O, in particular, the copper split vacancy. We demonstrated that the symmetry reduction due to this defect activates all zone center phonons of Cu_2O. The copper split vacancy is a strong perturbation of the crystal structure with a low formation energy and thus more likely to lift the Raman selection rules than defect activated phonons from different critical points of the Brillouin zone.

ACKNOLEGMENTS

We thank T. Buonassisi, M. Eickhoff, and B. K. Meyer for the provision of samples.

REFERENCES

1. W. H. Brattain, Rev. Mod. Phys. **23**, 203 (1951).
2. B. K. Meyer, A. Polity, D. Reppin, M. Becker, P. Hering, P. J. Klar, Th. Sander, C. Reindl, J. Benz, M. Eickhoff, C. Heiliger, M. Heinemann, J. Bläsing, A. Krost, S. Shokovets, C. Müller, C. Ronning, Phys. Status Solidi B **249**, 8 (2012).
3. M. Balkanski, M. Nusimovici, J. Reydellet, Solid State Commun. **7**, 815 (1969).
4. C. Carabatos and B. Prevot, Phys. Status Solidi B **44**, 701 (1971).
5. K. Huang, Zeitschrift für Physik **171**, 213 (1963).
6. J. Reydellet, M. Balkanski, D. Trivich, Phys. Status Solidi B **52**, 175 (1972).
7. A. Compaan and H. Z. Cummins, Phys. Rev. B **6**, 4753 (1972).

8. Y. Petroff, P. Y. Yu, Y. R. Shen, Phys. Rev. Lett. **29**, 1558 (1972).
9. P. Dawson, M. Hargreave, G. Wilkinson, J. Phys. Chem. Solids **34**, 2201 (1973).
10. P. F. Williams and S. P. S. Porto, Phys. Rev. B **8**, 1782 (1973).
11. A. Compaan, Solid State Commun. **16**, 293 (1975).
12. D. Powell, A. Compaan, J. R. Macdonald, R. A. Forman, Phys. Rev. B **12**, 20 (1975).
13. K. Reimann and K. Syassen, Phys. Rev. B **39**, 11113 (1989).
14. A. F. Wright and J. S. Nelson, J. of Appl. Phys. **92**, 5849 (2002).
15. M. Nolan and S. D. Elliott, Phys. Chem. Chem. Phys. **8**, 5350 (2006).
16. H. Raebiger, S. Lany, A. Zunger, Phys. Rev. B **76**, 045209 (2007).
17. D. O. Scanlon, B. J. Morgan, G. W. Watson, A. Walsh, Phys. Rev. Lett. **103**, 096405 (2009).

Mater. Res. Soc. Symp. Proc. Vol. 1633 © 2014 Materials Research Society
DOI: 10.1557/opl.2014.145

Characterization of mechanical, optical and structural properties of bismuth oxide thin films as a write-once medium for blue laser recording

Martyniuk M.[1], Baldwin D.[2], Jeffery R.[3], Silva K.K.M.B.D.[1], Woodward R.C.[4], Cliff J.[5], Krishnan R.N.[1], Dell J.M.[1], and Faraone L.[1]

[1] School of Electrical, Electronic and Computer Engineering, The University of Western Australia, 35 Stirling Hwy, Crawley, WA 6009, Australia

[2] 4Wave Inc., Sterling, VA, United States

[3] Panorama Synergy Ltd, Balcatta, WA, Australia

[4] School of Physics, The University of Western Australia, 35 Stirling Hwy, Crawley, WA 6009, Australia

[5] Centre for Microscopy, Characterisation and Analysis, The University of Western Australia, 35 Stirling Hwy, Crawley, WA 6009, Australia

ABSTRACT

We report on the preparation and characterization of crystalline bismuth oxide thin films via Biased Target Ion Beam Deposition method. A focused blue laser (405nm) is used to write an array of dots in the bismuth oxide thin film and demonstrate clear and circular recording marks in form of "bubbles" or "little volcanos" (FWHM ~500nm). Results indicate excellent static recording characteristics, writing sensitivity and contrast. The recording mechanism is investigated and is believed to be related to laser-induced morphology change.

INTRODUCTION

Bismuth oxide films have been demonstrated by a number of growth techniques, including reactive magnetron sputtering [1, 2, 3], Rapid thermal oxidation [4], air oxidized vacuum evaporation [5], Pulsed Laser deposition [6], and spray pyrolysis [7, 8]. Five separate crystalline polymorphs (α, β, γ, δ, ω) of Bi_2O_3 have been reported in the literature [8]. Additionally, two non-stoichiometric phases ($Bi_2O_{2.33}$ and $Bi_2O_{2.75}$) have also been reported [8]. The deposition technique and conditions have been shown to strongly affect the phase composition of the deposited Bismuth oxide material. The present work examines films of Bismuth oxide deposition using a relatively new technique, Biased Target Ion Beam Deposition (BTIBD).

A number of interesting optical and electrical properties have been demonstrated in bismuth oxide. These properties include a high optical nonlinearity in optical fiber [9, 10], good photosensitivity in the ultraviolet wavelengths [3], an optical bandgap varying from 2eV to 3.96eV depending on the deposition technology [3], and an electrical conductivity, dependent of the mode of deposition, which can vary over five orders of magnitude [6]. The property of importance to the present work is the demonstrated sensitivity to writing by a blue light, demonstrated previously by Jiang *et al.* [1].

Blu-ray media, players and recorders are fast replacing their DVD counterparts as the standard optical data storage format. According to the blu-ray specification, the media can contain up to 27 GB of data per recorded layer. To achieve this data density, the blu-ray player/recorder uses a 405 nm blue/violet laser, and a focusing lens with a numerical aperture of 0.85.

In order to achieve low-cost blu-ray disk media, a lot of attention is being paid to the use of organic dye-based recording layers [11]. However, little information exists on the stability and longevity of these dye-based recording media. Generally, dye-based media can be degraded by exposure to heat and sunlight, and the dye may be soluble in household organic solvents. If past experience with organic dye-based CD and DVD media can be applied to blu-ray media, these disks will likely not be suitable for long-term archival use. For archival use, the recording layer should ideally be a lot more stable than organic-dye based recording media of the past. It should be tolerant to environmental heat and light, and be resistant to household organic solvents. These requirements naturally point one to look at inorganic materials.

Previous studies have revealed a number of interesting optical properties of bismuth oxide, including a good sensitivity to writing by a blue laser, demonstrated by Jiang *et al.* [1]. The material used by Jiang *et al.* has been amorphous as deposited, and it has been speculated that it is not crystallization but, rather, bubble formation and change in grain size that causes the change in reflectivity [1].

The work presented here examines bismuth oxide grown by biased-target ion beam deposition (BTIBD) [12], and its potential as a blu-ray recording medium. This is a relatively new deposition technique, which uses a high-current, low-energy neutralized ion beam source to provide a broad, uniform flux of ions to a negatively biased sputter target. Because of this, the targets need not have magnetic fields associated with it, as in magnetron sputtering, and the sputtering ion energy (target bias voltage) in BTIBD can be reduced as low as desired without loss of ion/plasma generation, as would happen below ~ 400 Volts in magnetron sputtering. BTIBD is especially advantageous over magnetron sputtering for deposition of oxides by reactive sputtering of the metallic target using, e.g., O_2/Ar gas mixtures, because there is no magnetron "race track" region of intense ion bombardment on the surface of the target, so both ion bombardment and oxygen adsorption on the target surface are uniform and stable under all sputtering conditions and gas flow changes, after < 1 ms settling time. BTIBD widens the accessible process parameter ranges, and the technique is scalable to production substrate sizes and material growth rates. For the samples we have received from 4Wave, Inc. for this study, the kinetic energy of Bi species sputtered from the Bi metal target was reduced, and the process results in highly crystalline as-deposited material. Furthermore, the BTIBD bismuth oxide material is stoichiometric and of a single phase, rather than the mixed-phase material put down by other deposition techniques, and in particular as reported Jiang *et al.* [1].

THE EXPERIMENT

X-ray diffraction (XRD), optical transmission and nanoindentation were used to investigate the material properties of the deposited bismuth oxide.

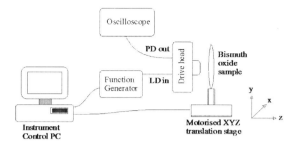

Fig. 1. Schematic view of the experimental setup for exposing bismuth oxide samples with blue laser.

Figure 1 schematically illustrates the experimental setup used to expose the Bismuth oxide material to blue light. Rather than building an expensive custom setup such as that of Gao *et al.* [13] we opted to use a relatively inexpensive commercial optical system already designed for exposing materials with blue light – the optical head removed from a blu-ray optical drive. This head contains all the optimized optics required for delivering the blue laser light to the sample, producing a 0.5 µm spot size. It also contains the return optics to collect the reflected light from the sample, and direct it to the photodiode used for detection. Measurements of the blue laser contained in the drive head showed it had a centre wavelength of 405 nm. The optical output from the drive head was measured to be 11 mW.

As we have no access to the electronics of the drive, the laser driver circuitry had to be custom built. Furthermore, the detector used in the blu-ray head contained internal circuitry and, prevented the extraction of the photodiode signal. For this reason, a third-party blue-enhanced photodiode was purchased, and a simple biasing circuit was built to drive this photodiode. The sample was smaller and easier to move than the blu-ray head. Therefore, the sample was mounted on a XYZ translation stage, built from three individual motorized translation stages. Each translation stage attached to the computer via a USB interface, and provided an ActiveX control for programmatic control.

A function generator is used to create electrical pulses, of the required amplitude and timing, to drive the modulation input of the laser diode driver circuit. This electrical modulation produces a pulsed output of the laser diode. The photoreceiver circuit output is directed to an oscilloscope for visualization of the signal reflected from the sample. In order to aid the visualization of the pulses on the oscilloscope screen, the synch output of the function generator is used to trigger the oscilloscope display. The function generator is remotely programmable via a USB interface, allowing PC control of the pulse characteristics, and of the triggering of the pulses. The programmable function generator, along with the computer controllable translation stages, facilitate the exposure of an accurate grid of points on the sample.

The control software for the sample exposure system was developed in the *LabVIEW*™ programming language from National Instruments Inc.

RESULTS AND ANALYSIS

The bismuth oxide layer of nominal thickness of 490 nm was deposited on a 1" diameter substrate of quartz using BTIBD. Nanoindentation experiments reported that the film was characterized by hardness and elastic modulus values of 5 GPa and 90 GPa, respectively. Optical transmission measurements indicated 80% relative average transmission for optical wavelengths between 500nm and 750nm. Figure 2 presents XRD data for the BTIBD as-grown bismuth oxide. The pattern obtained is compared to x-ray diffraction card ICSD 98-004-1764 and identifies as preferentially oriented β-Bi_2O_3. The sharp (021) peak, along with the absence of any other peaks apart from the (042) multiple, is indicative of highly oriented (more than 90% as shown by Rietveld refinement) tetragonal Bi_2O_3 with lattice spacing a=b=7.743Å and c=5.630Å. The XRD rocking curve corresponding to the (021) reflection was characterized by a full-width-half-maximum of 166 arcsec, confirming the preferential orientation of the as-deposited thin films of bismuth oxide.

The observation of as-sputtered, crystalline highly-oriented single phase bismuth oxide on an amorphous fused quartz substrate has not been reported in prior literature. In particular, Jiang et al. [1] obtain as-deposited material that shows mixed phases of tetragonal $Bi_2O_{2.75}$ and hexagonal Bi, which required annealing at 300°C for 30 minutes to transform into a mixture of face-centered cubic δ-Bi_2O_3 and hexagonal BiO (δ-Bi_2O_3 was thought to be the main component).

The deposited sample, with substrate, was held in a 1" optic mount for laser exposure. The translation stage was moved, in a x-y grid, with 5μm separation between steps. At each step, the function generator is triggered to generate a pulse of 700ns duration. In this way, a grid of

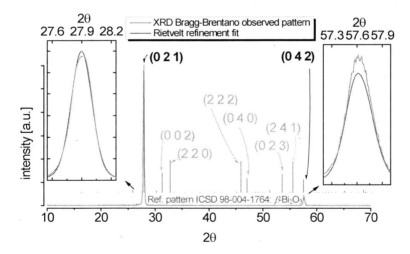

Fig. 2. X-ray diffraction data for BTIBD grown bismuth oxide sample

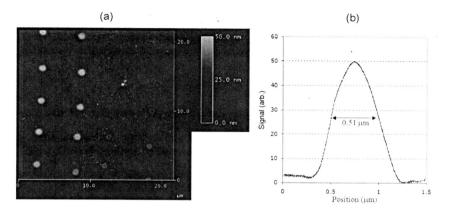

Fig. 3. (a) AFM image of exposed region, and (b) cross section through high-resolution scan of a single exposed "dot".

points is written to the sample. The region of exposure was subjected to atomic force microscopy (AFM) imaging. The AFM images of the exposed region, at two different spatial scales, are shown in Fig. 3. These images show a grid of points, with 5μm spacing. The individual structures are approximately 0.5 μm in width, and 50 nm in height. The dimmer points seen in Figure 3 (a) are likely the result of an overlapping out-of-focus previous exposure over this area.

The results indicate that BTIBD bismuth oxide is capable of supporting information encoding by blue light exposure. Since as-deposited BTIBD bismuth oxide is crystalline, the writing mechanism cannot be due to a simple amorphous to crystalline phase change. It has been suggested by Jiang et al.'s work [1] that writing mechanism could be associated with laser exposure induced transition of amorphous and/or multiphase BiO_x to a predominantly single phase δ-Bi_2O_3. However, our observations do not support this hypothesis as in our case the pre-exposed material is already predominantly single phase before laser exposure. Rather than an amorphous to crystalline phase change, the writing mechanism may still be due to bubble formation and grain-size changes. The shape of the exposed "dots" or "volcanos" does support the hypothesis that some form of bubble-formation is taking place.

CONCLUSIONS

As-grown crystalline bismuth oxide thin films were prepared via Biased Target Ion Beam Deposition method. The films as-deposited on amorphous fused quartz substrates are characterized by a crystalline material with a highly oriented (021) preferred tetragonal orientation. A focused blue laser (405nm) beam exposure of the bismuth oxide thin film demonstrates clear and circular recording marks in form of "bubbles" or "little volcanos" (FWHM ~500nm), indicating excellent static recording characteristics, writing sensitivity and contrast.

ACKNOWLEDGEMENTS

We acknowledge the support from the Australian Research Council, The Western Australian Node of the Australian National Fabrication Facility, and the Office of Science of the WA State Government.

REFERENCES

[1] Z. Jiang, Y. Geng, and D Gu, *Write-once medium with BiOx thin films for blue laser recording*, Chinese Opt. Lett., vol. 6, no 4, pp. 294 – 296, 2008.
[2] L. Leontie, M. Caraman, I. Evtodiev, E. Cuculescu, and A. Mija. *Optical properties of bismuth oxide thin films prepared by reactive d.c. magnetron sputtering onto p-GaSe (Cu)*, Phys. Stat. Sol. A, vol. 205, no. 8, pp. 2052 – 2056, 2008.
[3] L. Leontie, *Optical properties of bismuth oxide thin films prepared by reactive magnetron sputtering*, J. Optoelectronics Adv. Mat., Vol 8, No 3, pp. 1221 – 1224, Jun 2006.
[4] R. A. Ismail, *Characteristics of Bismuth trioxide film prepared by rapid thermal oxidation*, e-J. Surf. Sci. Nanotech. vol. 4, pp. 563 – 565, 2006.
[5] R.B. Patila, J.B. Yadava, R.K. Puria, and V. Purib, *Optical properties and adhesion of air oxidized vacuum evaporated bismuth thin films*, J. Phys. Chem. Solids, vol. 68, pp 665 – 669, 2007.
[6] L. Leontie, M. Caraman, A. Visinoiu, G.I. Rusu, *On the optical properties of bismuth oxide thin films prepared by pulsed laser deposition*, Thin Solid Films, vol. 473 pp. 230 – 235, 2005.
[7] V.V. Killedar, C.H. Bhosale, and C.D. Lokhande, *Characterization of Spray Deposited Bismuth Oxide Thin Films from Non-Aqueous Medium*, Tr. J. Phys., vol. 22, pp. 825 – 830, 1998.
[8] T.P. Gujar, V.R. Shinde, C.D. Lokhande, *Spray pyrolysed bismuth oxide thin films and their characterization*, Mat. Res. Bulletin, vol. 41, pp. 1558 – 1564, 2006.
[9] N. Sugimoto, T. Nagashima, T. Hasegawa, S. Ohara, K. Taira and K. Kikuchi, *Bismuth-based optical fiber with nonlinear coefficient of 1360 $W^{-1}km^{-1}$*, In proceedings of OFC/NFOEC, Los Angeles, California, Feb 22-27, 2004.
[10] F. Poletti, P. Petropoulos, N. G. R. Broderick and D. J. Richardson, *Design of Highly Nonlinear Bismuth-Oxide Holey Fibres with Zero Dispersion and Enhanced Brillouin Suppression*, In proceedings of Europ. Conf. Opt. Comms, Cannes, France, Sep 2006.
[11] T.-T. Hung, Y.-J. Lu, W.-Y. Liao, and C.-L. Huang, *Blue Violet Laser Write-Once Optical Disk With Coumarin Derivative Recording Layer*, IEEE Trans. Magnetics, vol. 43, no. 2, 2007.
[12] T. L. Hylton, B. Ciorneiu, D. A. Baldwin, O. Escorcia, J. Son, M. T. McClure, and G. Waters, *Thin film processing by biased target ion beam deposition*, IEEE Trans. Magnetics, vol. 36, no. 5, pp 2966-2971, Sep 2000.
[14] X. Gao, W. Xu, Fuxi Gan, F. Zhang, and F. Huang, *Design and construction of a static tester for blue ray optical storage*, Optik, vol. 117, pp 355–362, 2006.

Device Issues

High Performance IGZO TFTs with Modified Etch Stop Structure on Glass Substrates

Forough Mahmoudabadi[1], Ta-Ko Chuang[2], Jerry Ho Kung[3], and Miltiadis K. Hatalis[1]

[1]Display Research Laboratory, Lehigh University, Bethlehem, PA 18015, USA
[2]Corning Incorporated, Corning, NY 14831, USA
[3]Department of Electro Optical Engineering, National United University, Miaoli 36003, Taiwan

ABSTRACT

In this paper, we present fabrication and characterization of RF sputtered a-IGZO TFTs having a modified etch stopper structure with source/drain contact windows on glass wafers. The effect of annealing time and channel length on device performance in terms of mobility, on/off current ratio, average off current, threshold voltage, and sub threshold slope is reported.

INTRODUCTION

Amorphous oxide semiconductors and in particular, amorphous indium-gallium-zinc oxide (a-IGZO) thin films are gaining considerable interest for display and flexible electronics applications. At present, thin film transistor technologies based on hydrogenated amorphous silicon (a-Si:H) and polycrystalline silicon (poly-Si) are widely utilized in production of flat panel displays. However, the low mobility in a-Si:H TFTs limits their applications in large high resolution displays. Poor stability of a-Si:H TFTs also limits their use in AMOLED display applications. The cost of the poly-Si TFT is considered to be relatively high. Furthermore, the grain boundaries present within poly-Si deteriorates device performance uniformity. These are some reasons for the ever increasing interest for developing metal oxide semiconductors like a-IGZO TFTs. Studies have shown that a-IGZO are superior to a-Si:H and organic semiconductors in terms of stability. Other reported advantages of amorphous metal oxides such as a-IGZO include good uniformity, high mobility, and feasibility of room temperature processing; these advantages make them good alternatives to both silicon based and organic semiconductors for display and other large area electronic applications [1-3].

This paper reports on characteristics of a-IGZO TFTs having a modified etch stop structure and investigates the effect of channel length on TFT performance. The bottom gate TFTs in this study are fabricated on glass substrates.

EXPERIMENTAL

Figure 1 shows a photograph and a cross sectional view of the TFT structure. The bottom-gate TFTs were fabricated by first depositing 140 nm of Aluminum doped with Nd (AlNd) onto 150 mm Corning LOTUS glass substrates and patterning the aluminum film by lift-off in order to form gate electrodes. A 120-nm-thick SiO_2 layer was then deposited by PECVD to serve as the gate dielectric. RF sputtered 50 nm of IGZO thin film formed the active layer of the TFTs. Deposition was done at 150 W from a 150 mm target in an argon/oxygen ambient at a deposition pressure of 6 mtorr. A 50 nm RF-sputtered SiO_2 film served as the first passivation layer (active passivation layer shown in Figure 1(b)) which protected the surface of the underlying IGZO film during subsequent processing steps. The SiO_2 and IGZO layers were patterned by dry and wet etching processes, respectively. Then, a second 50 nm SiO_2 passivation

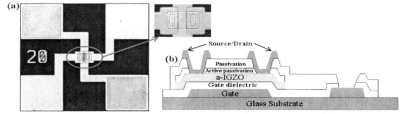

Figure 1. (a) Top-view photograph of fabricated TFT (b) schematic cross-sectional view of TFT structure

layer was deposited in order to protect the edges of the IGZO film and provide additional protection on its top surface. Afterwards, two different patterning and dry etching steps were performed in order to form the contact windows on top of the source and drain regions as well as to open the gate pads as shown in Figure 1. Finally, a double layer of Mo/AlNd electrode formed the source and drain electrodes through a lift-off process. The device fabrication was then completed by a thermal annealing at 300 °C in nitrogen ambient; various annealing times were investigated. TFTs having a variety of geometries in terms of channel length and width were fabricated on each glass wafer; a total of twelve wafers were used. From each wafer, a total of 41 dies were characterized at room temperature under dark condition using an HP 4145A semiconductor parameter analyzer and an automated probe station.

RESULTS AND DISCUSSIONS

Typical transfer and output characteristics for a-IGZO TFTs are shown in Figure 2. Device transfer characteristics and gate leakage current were measured at V_{DS} = +0.1 V and +10 V for a V_{GS} in the range of -10 to +20 V. Mobility was extracted from the maximum transconductance while the threshold voltage was obtained using the linear extrapolation method from the characteristics obtained at V_{DS} = 0.1 V using the following equation:

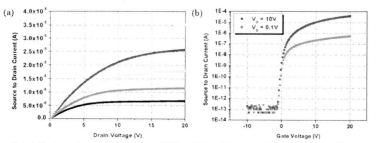

Figure 2. (a) Output characteristics of a TFT with L= 16 µm and W=10 µm for V_{GS} values starting from 8V and with increments of 3 V (b) Transfer characteristics of the same device for V_{DS} biases of 0.1 V and 10 V.

$$I_{DS} = \frac{W}{L} \mu_{FE} C_{OX} V_{DS} (V_{GS} - V_{th}) \quad (1)$$

where μ_{FE} is field effect mobility, C_{OX} is the dielectric capacitance per unit area, and W is the channel width. Channel length, L, is defined as the distance between source and drain contacts (figure 1(b)). TFTs with length of 16 μm and width of 10 μm showed an average field effect mobility of 15.5 cm^2/Vs (with standard deviation of 2.5) and threshold voltage of 4.8 volts (with standard deviation of 1) after a total of three hours annealing in nitrogen ambient.

It was reported [4] that thermal annealing reduces the density of shallow localized states beneath the conduction band minimum and thereby improves electron transport in amorphous metal oxide semiconductors. In our case, increasing the annealing time improved both the electrical characteristics of TFTs of all geometries and also increased the number of functional devices. Figure 3 displays the TFT fabrication yield as a function of annealing time and channel length. All TFTs before the first anneal were non-functional (in a non-conductive phase). After a total of three hours, almost all short channel length (i.e. L < 20 μm) and around half of the long channel length (i.e. L > 20 μm) devices became functional. Further annealing (results not shown) improved the performance of remaining long length TFTs to the point to show good characteristics. However, the performance of short channel TFTs in terms of mobility and threshold voltage was degraded. Therefore, a maximum of three hour annealing was considered as the standard annealing time in our experiment.

As figure 4 shows the long channel length devices in a thicker IGZO layer (figure 4(a)), located around the center of each glass wafer, required more annealing time than devices fabricated in a thinner IGZO layer located around the edge (figure 4(b)). The IGZO thickness variation between the center and the edge were ~20%. Figures 5(a) and 5(b) display electrical parameters of TFTs having the same width and thickness as a function of channel length. As it is seen, shorter channel TFTs exhibit better performance. It should be noted that the trend in TFT operation illustrated in figure 4 and figure 5 was repeated in all devices (around 6000 TFTs) fabricated on twelve wafers. While the TFT parameters were slightly different from wafer to wafer due to process variations, all TFTs on each wafer showed similar behavior in terms of dependency of electrical parameters on channel length. To investigate this phenomenon further, the parasitic source-to-drain resistance and the effective channel length were evaluated using the channel resistance method [5] with TFTs having different lengths. The TFT total resistance is

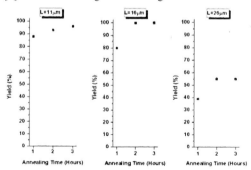

Figure 3. Effect of annealing time and channel length on TFT yield

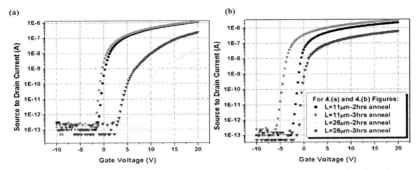

Figure 4. Transfer characteristics as a function of channel length and annealing time for devices located at (a) center and (b) edge of the substrate. Drain voltage and TFT width are 0.1V and 20 μm, respectively.

defined by the following equation:

$$R_{tot} = \frac{V_{DS}}{I_{DS}} = R_{SD} + \frac{L-\Delta L}{\mu_{eff} C_{OX} W (V_{GS}-V_{th})} \quad (2)$$

where L is the physical channel length (distance between source and drain contacts), and effective length (L_{eff}) is defined by $L_{eff} = L-\Delta L$. Here, ΔL is an apparent channel length reduction that can be obtained from the intersection point of the R_{tot}-length straight lines, as shown in the Figure 6. μ_{eff} is the effective field effect mobility that is obtained by replacing L with L_{eff}, and C_{OX} is the gate dielectric capacitance per unit area. The width-normalized R_{SD} ($R_{SD}W$) is evaluated to be 51 Ω.cm. This observation confirms that low ohmic contacts are formed in the

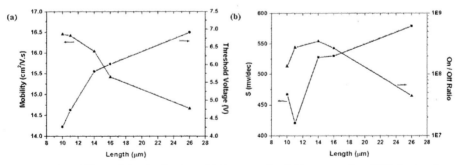

Figure 5. (a) Mobility and threshold voltage (b) S value and on/off current ratio of TFTs having the same width and thickness as a function of channel length

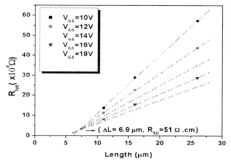

Figure 6. R_{tot} vs channel length as a function of V_{GS} for TFTs with W=20 μm

present process. The length reduction of ΔL = 6.9 μm is consistent with that reported in literature for metal oxide TFTs with Mo electrode [6, 7]. Based on literature, we consider two possible explanations for the channel length reduction and the channel length dependency of the TFT parameters in our experiment.

The channel length reduction might be attributed to diffusion of Mo into the a-IGZO layer [6]. If this happens, Mo diffusion may decrease the effective channel length which will result in higher output current. It has also been reported that in indium-zinc oxide TFTs with molybdenum source and drain electrodes, the threshold voltage has a dependence on channel length and this was attributed to the reduction of channel length [6] due to molybdenum diffusion in the IZO.

The dependency of the threshold voltage on channel length in our IGZO TFTs might be attributed to oxygen vacancies produced at the IGZO/Mo interface. It has been known that electrical properties of metal oxides can be changed by modifying the oxygen content of the film because the conduction electrons originate at oxygen vacancies [8-10]. It was also reported [11-15] that in a-IGZO TFTs with metal electrodes; metal atoms can diffuse into a-IGZO and change the thermodynamic equilibrium of a-IGZO. In a-IGZO TFTs with Ti electrodes [14], Ti reacted with oxygen from the IGZO layer and generated a thin Ti oxide layer at the IGZO/Ti interface. As a result, a highly conductive, oxygen deficient region was formed in the IGZO layer under Ti electrode and a large negative shift in threshold voltage was observed due to oxygen vacancies in a-IGZO layer. A similar trend with IGZO TFTs having Al electrodes was reported [15]. Based on Gibbs free energy data obtained from literature [11, 16], formation of MoO_3 (ΔG = -591 KJ/mol) is thermodynamically more favored compared to In_2O_3 (ΔG = -490 KJ/mol) at 300 ℃ (annealing temperature in our experiment). Higher concentration of oxygen vacancies at the Mo/IZO interface of TFTs with Mo electrodes was also reported [6].

Thereby, based on these reports and since Gibbs free energy calculation suggests that Mo oxidation is favorable even at low temperature [17], we consider the same phenomenon might be happened in this experiment. After diffusion of Mo into a-IGZO semiconductor, Mo atoms were oxidized and caused reduction of indium oxide resulting in an increase of semiconductor conductivity. Since the concentration of oxygen vacancies in the IGZO film induced by electrodes is higher in shorter channel length TFTs, these devices will exhibit a lower threshold voltage. As shown in Fig. 4, a dependency between IGZO layer thickness and performance of

TFTs was also observed in TFTs having a long channel length. TFTs in thinner IGZO located around the edge of each wafer displayed superior characteristics in a shorter time compared to TFTs in thicker IGZO located at the center of a wafer. Due to length limitation of this paper, detailed on this observation and results of our ongoing investigation will be reported in a later publication [18].

SUMMARY

In conclusion, we have demonstrated bottom-gate a-IGZO TFTs having a modified etch-stop structure with very good performance: an average field effect mobility of ~ 15.5 cm^2/V.s, threshold voltage of ~ 4.8 V, sub threshold slope of ~ 450 mV decade^{-1}, on-to-off current ratio of ~ > 10^8, and average off current of ~ 10^{-13} for TFTs with channel length and width of 16 μm and 10 μm, respectively. Annealing time resulted in a small improvement in device characteristics but had a big effect on the number of functional devices. A dependence of device performance on the channel length was observed and it was attributed to the diffusion and reaction of the molybdenum source and drain electrode with the IGZO channel layer.

ACKNOWLEDGEMENT

The work at Lehigh University was supported by a grant from Corning Incorporated.

REFERENCES

1. K. Nomura, et al. *Nature* 432, no. 7016 (2004).
2. W. Lim, et al. *Journal of The Electrochemical Society* 155, no. 6 (2008).
3. J. S. Park, et al. *Thin Solid Films* 520, no. 6 (2012)
4. H. Hosono, et al. *Journal of non-crystalline solids* 354, no. 19 (2008).
5. S. E. Laux. *Electron Devices, IEEE Transactions on* 31, no. 9 (1984).
6. L. Lan, M. Xu, J. Peng, H. Xu, M. Li, D. Luo, J. Zou, H. Tao, L. Wang, and R. Yao. *Journal of Applied Physics* 110, no. 10 (2011).
7. J. S. Park, T. S. Kim, K. S. Son, E. Lee, J. S. Jung, K. H. Lee, W. J. Maeng et al. *Applied Physics Letters* 97, no. 16 (2010).
8. J. Yao, N. Xu, S. Deng, J. Chen, J. She, H. D. Shieh, Po-Tsun Liu, and Yi-Pai Huang. *Electron Devices, IEEE Transactions on* 58, no. 4 (2011).
9. D. Kang, et al. *Applied physics letters* 90, no. 19 (2007).
10. S. Honda, et al. *Journal of Vacuum Science & Technology A: Vacuum, Surfaces, and Films* 13, no. 3 (1995).
11. S. H. Choi, W. S. Jung, and J. H. Park. *Applied Physics Letters* 101 (2012).
12. K. H. Choi, and H. K. Kim. *Applied Physics Letters* 102, no. 5 (2013).
13. K. Park, et al. *Journal of Materials Research* 25, no. 02 (2010).
14. T. Arai, et al. *SID Symposium Digest of Technical Papers*, vol. 41, no. 1, pp. 1033-1036. Blackwell Publishing Ltd, 2010.
15. J. R. Yim, et al. *Japanese Journal of Applied Physics* 51, no. 1R (2012)
16. D. R. Stull and H. Prophet. *JANAF thermochemical tables*. no. NSRDS-NBS-37. National Standard Reference Data System, 1971.
17. I. Barin, et al. *Thermochemical properties of inorganic substances: supplement*. Vol. 380. Berlin: Springer-Verlag, 1977.
18. F. Mahmoudabadi, et al. (Unpublished).

Mater. Res. Soc. Symp. Proc. Vol. 1633 © 2014 Materials Research Society
DOI: 10.1557/opl.2014.117

Amorphous zinc-tin oxide thin films fabricated by pulsed laser deposition at room temperature

P. Schlupp, H. von Wenckstern and M. Grundmann
Universität Leipzig, Fakultät für Physik und Geowissenschaften, Institut für Experimentelle Physik II, Linnéstrasse 5, 04103 Leipzig, Germany

ABSTRACT

For a cost-efficient fabrication of homogeneous oxide thin films the usage of amorphous materials is favorable. They can be deposited at room temperature (RT) and represent an interesting alternative to amorphous silicon in electronics. Zinc-tin oxide is a promising n-type channel material for thin film transistors and consists of abundant elements, only, in contrast to the well-explored indium gallium zinc oxide. Here, the electrical and optical properties of room temperature deposited ZTO thin films are discussed. These films were fabricated via pulsed-laser deposition on glass substrates by ablating a ceramic target composed of ZnO and SnO_2 in a 1:2 ratio. The resistivity has been controlled over seven orders of magnitude via the oxygen growth pressure. Further, the optical transmittance tends to be higher for higher oxygen growth pressures.

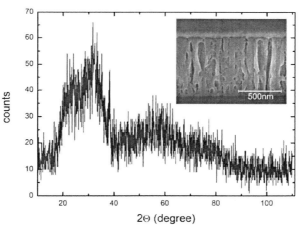

Fig. 1: XRD pattern of an amorphous ZTO thin film. The inset shows a SEM image of a ZTO thin film grown at $p_{O2} = 10$ Pa.

INTRODUCTION

Amorphous semiconducting oxides are promising candidates for functional and cost-efficient thin films. Room temperature fabrication is additionally compatible with plastic substrates and makes flexible device applications feasible. Zinc-tin oxide (ZTO) is a promising amorphous *n*-type channel material[1,2,3]. Here, we present electrical and optical properties of ZTO thin films deposited by pulsed-laser deposition (PLD) at room temperature. Hall-effect measurements carried out with the van-der-Pauw method and transmittance measurements in the visible and UV spectral range were performed.

EXPERIMENT

Amorphous ZTO thin films were fabricated in three different growth chambers of different geometry by PLD. For target ablation we used a KrF excimer laser (λ=248 nm). The ceramic targets consist of 66.7 at.% SnO_2 and 33.3 at.% ZnO. The oxygen partial pressure during the deposition was varied from 0.03 Pa to 10 Pa. All thin films deposited on glass substrates (Corning 1737, 10×10 mm^2) have a thickness between 500 nm and 1 μm. For the electrical characterization Hall-effect measurements were performed using a home-built set up, which is described in Ref. [4]. As ohmic contacts gold layers were sputtered onto the corners of the thin films. For X-ray diffraction (XRD) measurements a Philips X'pert with a Bragg-Brentano goniometer and Cu K_α radiation was used. Transmission measurements were performed using a Perkin Ellmer Lambda 40 UV/VIS spectrometer operating from 190 nm to 1100 nm with a step width of 0.1 nm and a measurement rate of 240 nm/min.

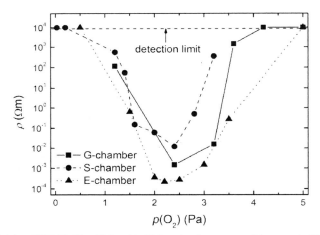

Fig. 2: Resistivity of ZTO thin films fabricated by PLD against the oxygen partial pressure in different growth chambers.

DISCUSSION

The ZTO thin films were fabricated in three different chambers (labeled E-, S- and G-chamber from now on). To confirm the amorphous structure of the ZTO thin films XRD measurements were performed. In Fig. 1 a typical XRD pattern is shown. The broad peak around 30 degree is typically observed for amorphous films of ZTO on glass substrates[1,5].

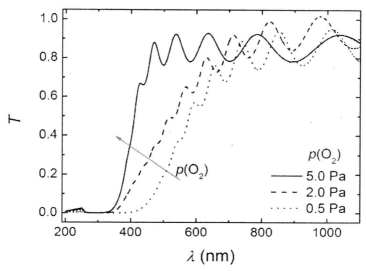

Fig. 3: Transmission of the about 1 μm thick ZTO films depending on the wavelength. Values higher than one are due to small variations between the substrate transmittance and that of a reference substrate used for calculation.

The Au contacts on the corners of the substrate cause the peak at 38 degree being due to the reflection on the (111)-planes. In conclusion the ZTO thin films are x-ray amorphous as expected for room temperature fabrication. In Fig. 2 the dependence of resistivity on the oxygen partial pressure is depicted for all sample series. Due to the differences in growth chamber geometry, the fabricated film series show weak variations in their electrical properties. Nevertheless all ZTO thin films fabricated at high partial pressure $p_{O2} > 4$ Pa and at low partial pressure $p_{O2} < 0.5$ Pa show resistivity above the detection limit of 10^4 Ωm. In the intermediate range the resistivity decreases by more than seven orders of magnitude to values as low as 2×10^{-4} Ωm. This high conductivity is due to the well-known non-directionality of the spherically symmetric ns-orbitals of Sn^{4+} and the Zn^{2+} ions forming the conduction band minimum[6]. Using 10 Pa as oxygen partial pressure during the fabrication the films are no longer compact. The inset of Fig. 1 shows a SEM image of such a ZTO thin film. A lot of vertically arranged voids are observed. These are not observed in thin films fabricated at $p_{O2} \leq 5$ Pa. But maybe they still start to form and with that the resistivity increases.

Another reason for the high resistivity could be nano-sized crystalline clusters. Samples annealed for two hours in air show a polycrystalline structure but the high conductivity vanishes during the annealing process (not shown here). Both would be possible explanations for the increasing resistivity in the films grown at pressures around $p_{O2} = 5$ Pa. For the films with the lowest resistivity electron density of 1×10^{19} cm^{-3} and Hall-effect mobility as high as 12.7 cm^2/Vs were observed. Electrons are the majority carriers as verified by Seebeck-effect measurements. The lowest free electron concentration we determined unambiguously was 1×10^{16} cm^{-3}. The possibility of tuning the electric properties, especially the charge carrier density, is very important for application in electronic devices like diodes and transistors.

Similar to the resistivity also the transmittance depends on the growth pressure, which is shown in Fig. 3. The higher p_{O2} the higher is the transmittance in the visible range. For growth pressures higher than 3 Pa a transmission of nearly 80% is achieved in the visible spectral range.

CONCLUSIONS

In conclusion it was demonstrated that the resistivity and the transmittance of zinc-tin oxide PLD thin films are easily controllable by the oxygen growth pressure. Hall-effect mobility as high as 12.7 cm^2/Vs and electron density of 1×10^{19} cm^{-3} were observed. In the visible spectral range the 1 μm thick films show a transmittance of 80%. Therefore ZTO is a good candidate for electronics fabricated at RT even on flexible substrates.

ACKNOWLEDGMENTS

The authors thank J. Lenzner (Universität Leipzig) for recording SEM images. P. S. is supported by Leipzig School of Natural Science (BuildMoNa). This work was supported by the ESF(SAB 100124929).

REFERENCES

[1] Madambi K. Jayaraj, Kachirayil J. Saji, Kenji Nomura, Toshio Kamiya and Hideo Hosono. J. Vac. Sci. Technol. **B 26**, 495 (2008)

[2] H. Q. Chiang, J. F. Wager, R. L. Hofman, J. Jeong, and D. Keszler, Appl. Phys. Lett. **86**, 013503 (2004)

[3] Jaeyeong Heo, Sang Bok Kim, and Roy G. Gordon, Appl. Phys. Lett. **101**, 113507 (2012)

[4] H. von Wenckstern, M. Brandt, G. Zimmermann, J. Lenzner, M. Lorenz and M. Grundmann. Mater. Res. Soc. Symp. Proc. **957**, 0957-K03-02 (2007)

[5] D. L. Young, D. L. Williamson and T. J. Coutts. J. Appl. Phys. **91**, 1464 (2002).

[6] H. Hosono, N. Kikuchi, N. Ueda and H. Kawazoe, J. Non-Cryst. Solids **198-200**, 165 (1996)

Solution Processed Resistive Random Access Memory Devices for Transparent Solid-State Circuit Systems

Yiran Wang, Bing Chen, Dong Liu, Bin Gao, Lifeng Liu*, Xiaoyan Liu, Jinfeng Kang*

Institute of Microelectronics, Peking University, Beijing 100871, China

*Email: lfliu@pku.edu.cn; kangjf@pku.edu.cn;

ABSTRACT

A solution-processed method is developed to fabricate fully transparent resistive random access memory (RRAM) devices with a configuration of FTO/ZrO2/ITO, where the zirconium dioxide (ZrO2) layer was firstly deposited on fluorine tin oxide (FTO) substrate by sol-gel and then indium tin oxide (ITO) films were deposited on ZrO2 layer by sol-gel as the top electrodes.The solution processed FTO/ZrO2/ITO based RRAM devices show the fully transparency and excellent bipolar resistance switching behaviors. The resistance ratio between high and low resistance states was more than 10, and more than 100 switching cycles and good data retention and multilevel resistive switching have been demonstrated.

INTRODUCTION

Transparent solid-state circuit systems have been extensively studied as the promising technology for the application in large area electronic systems by stacking clear display, electronic paper or other transparent devices into clear spaces[1-3]. As one of the critical devices of the circuit systems, the transparent memory devices are needed but it is difficult to realize by the traditional memory technology. Oxide-based resistive random access memory (RRAM), which presents the excellent device performances and data storage functions, shows the great potential in transparent memory application. Although the transparent oxide thin films could be deposited by sputtering, chemical vapor deposition, atomic layer deposition, and pulsed laser deposition, the solution-processed technology[4] shows the great advantages for the low-cost and large-area film application.

In this work, a solution-processed method is developed to fabricate fully transparent RRAM (TRRAM) devices with a configuration of FTO/ZrO$_2$/ITO, where the zirconium dioxide (ZrO$_2$) layer was firstly deposited on fluorine tin oxide (FTO) substrate by sol-gel and then indium tin oxide (ITO) films were deposited on ZrO$_2$ layer by sol-gel as the top electrodes. Layers including oxide layer and top layer were fabricated by spin-coating using sol-gel method . The resistive switching behaviors of the devices are investigated.

EXPERIMENT

Figure 1. shows the configuration of our TRRAM device on glass substrate. 15-nm-thick ZrO2 films were fabricated on FTO/glass substrates by sol-gel process. For the active layer, as shown in figure 2., zirconium n-butoxide [Zr(OC$_4$H$_9$)$_4$] ,80wt. % in n-butanol , was used as precursor material[5]. Ethanol was selected as solvent. Acetylacetone(C$_5$H$_8$O$_2$) and acetic acid (CH$_3$COOH) were chosen as reagents in stabilizing the precursor. Usually, the mol ratio of zirconium butoxide : anhydrous ethanol : acetic acid is 1: 16 : 4 in a batch[6]. After mixing zirconium butoxide with ethanol, acetylacetone and acetic acid were added. The solution was stirred for more than 10 minutes until it became clear. Active layer was deposited by spin-coating at 3500 rpm for 30s, and then transferred on hot plate at 150°C for 5 min.

Figure 3. shows the fabrication of the top layer. Indium nitrate [$In(NO_3)_3·4.5H_2O$] was dissolved in anhydrous ethanol, which acted as solvent and mixed with acetylacetone. Meanwhile, tin chloride [$SnCl·4.5H_2O$] was dissolved in anhydrous ethanol[7]. Two solutions were mixed after each had been stirred for 5 min. Typically, the mol ratio of acetylacetone to indium is 3, and indium to tin is 10. After patterning, the top layer was deposited by spin-coating at 3000 rpm for 30s on active layer. The thickness of ITO layer is around 20nm and the sheet resistance is approximately 3.5 kΩ/square. Then devices were annealed at 500 °C for 2h in air ambient. The current-voltage performance of devices were measured by KEITHLEY4200 semiconductor parameter analyzer.

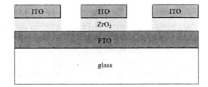

Figure 1. Structure of the prepared solution-processed ITO/ZrO_2/FTO TRRAM

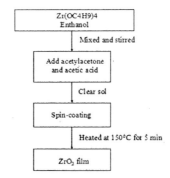

Figure 2. Preparation of ZrO_2 film by sol-gel process

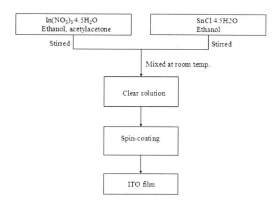

Figure 3. Preparation of ITO film by sol-gel process

RESULTS and DISCUSSION

Figure 4. shows the bipolar resistance switching behaviour of this solution processed FTO/ZrO$_2$/ITO based RRAM device. DC voltage sweep mode was set from 0V → 3V →0V→-3V→0V. 100 cycles had been tested, and the curves of 1st 10th, and 100th revealed the typical resistive switching characteristics. No forming process was observed. This may be attributed to the Vo defects already existed in ZrO$_2$ film as electrons hopping between oxygen vacancies (Vo) has been widely accepted as conducting mechanism in transition metal oxide[8].

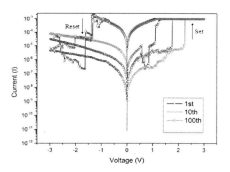

Figure 4. Typical bipolar resistive switching I-V curve of the solution processed ITO/ZrO$_2$/FTO TRRAM.

As for a bipolar RRAM, resistive switching (RS) happens when positive voltage sweeping was applied, which leads to a switch from high resistance state (HRS) to low resistance state

(LRS) at a SET voltage (V_{set}). After that, when voltage swept back to negative side, the LRS was switched back to HRS at a RESET voltage (V_{reset}). During SET process, in order to prevent high current in damaging the device, compliance current was fixed at 1 mA, Figure 5. shows the distribution characteristics of V_{set} and V_{reset} after switching for 100 times. The SET voltage varied from 0.9V to 2.7V. The standard deviation (σ_{Vset}) was 0.4V and the arithmetic mean value (μ_{Vset}) was 1.7V, which is less than 2V. By applying the voltage back to the negative side, the device switched from low resistance state (LRS) back to high resistance state (HRS) .The standard deviation (σ_{Vreset}) was 0.3V and the arithmetic mean value (μ_{Vreset}) was -1.3V. Programming current less than 1mA was demonstrated.

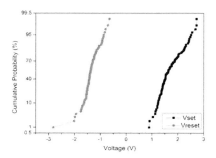

Figure 5. V_{set} and V_{reset} distribution characteristics of ITO/ZrO$_2$/FTO TRRAM after 100 cyclings.

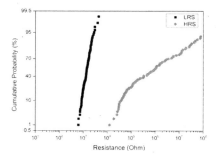

Figure 6. Resistance distribution characteristics of ITO/ZrO$_2$/FTO TRRAM after 100 cyclings. The on/off ratio was more than 10.

For nonvolatile memory devices, sufficient memory window (10 times) is satisfied. In our devices, as shown in figure 6., the resistance ratio between high and low resistance states read at 0.1V under room temperature was more than 10. After switching for more than 100 cycles, no obvious degradation of on/off ratio was observed. Resistance states retention characteristics of

HRS and LRS have been measured at room temperature and data retention over 1000s have been observed, as shown in Figure 7.

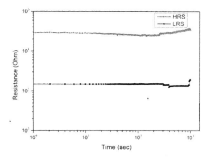

Figure 7. Retention characteristics for HRS and LRS at room temperature

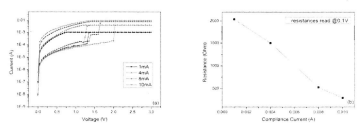

Figure 8. (a) I-V curves of ITO/ZrO_2/FTO TRRAM under various current compliances during SET process.(b) LRS resistances under different current compliances read at 0.1V

Multilevel resistive switching behaviors were also investigated. Figure 8. (a) and figure 8. (b) show the I-V characteristics for positive voltage sweeps and the set current compliance dependence of LRS resistance for the ITO/ZrO2/FTO TRRAM device, respectively. As shown in figure 8., by applying different current compliances during SET process[9], different LRS resistances were observed, which maintains potential in multilevel applications of this device.

CONCLUSIONS

In conclusion, fully transparent RRAM devices were fabricated under the development of solution processed method. Layers including top electrode and ZrO_2 were deposited by spin-coating using sol-gel method. The FTO/ZrO_2/ITO based RRAM devices show the fully transparency and excellent bipolar resistance switching behaviors. No forming process, low operation voltage less than 2V and programming current less than 1mA have been observed. The resistance ratio between HRS and LRS was more than 10, and more than 100 switching cycles, good data retention and multilevel resistive switching characteristics have been demonstrated. Since the solution-processed technology shows great advantages for the low-cost and large-area film manufacturing, our fully transparent RRAM devices have great potential in future RRAM applications, and thus for the application in large area electronic system.

ACKNOWLEDGMENTS

This work was supported in part by 973 and NSFC (Nos. 2011CBA00600, 2010CB934203, 61376084, 60925015).

REFERENCES

[1] J. F. Wager, *Science*, Vol. 300 No.5623, p. 1245, (May 2003).
[2] J. Yao, J. Lin, Y.H. Dai, G.D. Ruan, Z. Yan, L. Li, L. Zhong, D. Natelson, J.M. Tour, *Nature Communications*, Vol.3, No.1101, (Oct 2012).
[3] Seo, J. W., Park, J. W., Lim, K. S., Yang, J. H., & Kang, S. J., *Applied Physics Letters*, 93(22), 223505-223505, (2008).
[4] Kim, A., Song, K., Kim, Y., & Moon, J. ,*ACS applied materials & interfaces*, 3(11), 4525-4530, (2011).
[5] Bing, S., Li-Feng, L., De-Dong, H., Yi, W., Xiao-Yan, L., Ru-Qi, H., & Jin-Feng, K., *Chinese Physics Letters*, 25(6), 2187, (2008).
[6] Wu, Jeffrey Chi-Sheng, and Li-Chuen Cheng., *Journal of Membrane Science*,167.2: 253-261, (2000).
[7] Su, C., Sheu, T. K., Chang, Y. T., Wan, M. A., Feng, M. C., & Hung, W. C., *Synthetic Metals*, 153(1-3), 9-12, (2005).
[8] Zhang, H., Gao, B., Sun, B., Chen, G., Zeng, L., Liu, L.,Liu, X., Lu, J., Han, R., Kang, J., Yu, B., *Applied Physics Letters*, 96(12), 123502-123502, (2010).
[9]Wang, S. Y., Huang, C. W., Lee, D. Y., Tseng, T. Y., & Chang, T. C., *Journal of Applied Physics*, *108*, 114110, (2010).

Structural and Electrical Characteristics of Ternary Oxide SmGdO$_3$ for Logic and Memory Devices

Yogesh Sharma, Pankaj Misra and Ram S. Katiyar
Department of Physics, University of Puerto Rico, PR-00936-8377, USA

ABSTRACT

We report on the structural and electrical characteristics of bulk and thin film of ternary oxide SmGdO$_3$. Bulk sample of SmGdO$_3$ was prepared by pelletizing and sintering the calcined mixture of predetermined amount of Sm$_2$O$_3$ and Gd$_2$O$_3$ powders. The crystalline structure of the sample was studied by X-ray diffraction measurements and Raman spectroscopy. Capacitance and leakage current measurements on bulk sample revealed a high and linear dielectric constant of ~ 19 with low dielectric loss and leakage current which is suitable for gate dielectric application in CMOS logic devices and high-k MIM capacitors. In addition, the non-volatile resistive memory switching phenomenon was studied in thin films of SmGdO$_3$ which were deposited by pulsed laser deposition using sintered pellet of SmGdO$_3$ as target. Commercially available Pt/TiO$_2$/SiO$_2$/(100) Si was used as substrate and top Pt electrode of lateral dimension 40×40μm^2 were deposited by sputtering to construct Pt/SmGdO3/Pt MIM devices. After initial forming process which occurred at comparatively higher voltage, the Pt/SmGdO$_3$/Pt devices showed repeatable unipolar switching between high and low resistance states with low and well defined switching voltages. These properties indicate suitability of this material for the emerging logic and memory device applications.

INTRODUCTON

The lanthanide sesquioxides (Ln$_2$O$_3$) have been widely investigated because of their interesting optical, electronic and magnetic properties [1]. Among these, some of the Ln$_2$O$_3$ compounds have also emerged as potential high-k gate dielectric material for Metal-Oxide-Semiconductor (MOS) and Resistive Random Access Memory (ReRAM) device applications [2,3]. Most of the sesquioxides which exhibit high dielectric constant show high hygroscopic nature which deteriorates their ability in terms of high-k gate dielectric applications [4]. One of the Ln$_2$O$_3$ compounds, Lanthanum oxides (La$_2$O$_3$) which has high dielectric constant of ~ 27, is not stable in air and very hygroscopic to form hydroxide [4]. Interlanthanide oxides, also called as ternary oxides could be used to prevent the aforementioned problems of sesquioxides. The ternary oxides of La$_2$O$_3$ such as LaGdO$_3$ and LaErO$_3$ have been shown to exhibit excellent properties as compared to their constituent sesquioxides [5, 6]. However, it has been seen that because of very hygroscopic nature of Lanthanum oxides (La$_2$O$_3$), even ternary oxides encounter stability issue in gate-dielectric applications [5]. Therefore, it is imperative to investigate the Lanthanum free ternary high-k dielectric materials to overcome the hygroscopic issue in terms of gate-dielectric and ReRAM applications.

In this report we synthesised a Lanthanum-free novel high-k ternary oxide SmGdO$_3$ (SGO) and explored its structural and electrical properties in bulk and thin film forms, for the logic and memory device applications. The variation of real part of dielectric permittivity of bulk samples with frequency and the Capacitance versus Votlage (C-V) characteristics demonstrated the linear dielectric behaviour of SGO with the dielectric constant of ~19. The Current versus voltage (I-V) characteristic showed the hysteretic behaviour. Further, a unipolar resistive switching has been observed in the as fabricated thin film device Pt/SGO/Pt in metal-insulator-metal (MIM) stack, which exhibited excellent memory switching parameters in terms of ReRAM applications.

EXPERIMENTAL DETAILS

The Single phase polycrystalline SGO is fabricated by solid state reaction method. High purity cation oxides Sm_2O_3 (99.99%) and Gd_2O_3 (99.99%) were thoroughly mixed in a molar ratio of 1:1 to achieve the cation stoichiometry of SGO and heated in air at 1200 °C for 12 hours in high density Al_2O_3 (Alumina) crucible. The reacted product was grounded, compacted, and reheated at 1350 °C for 10 hours to ensure its homogeneity and single phase formation. The final powder was compacted into pellets at 5 ton load and sintered at 1400 °C for 10 hours. For I-V characteristics and dielectric measurements, the metal-insulator-metal (MIM) stack has been fabricated with SGO pellets of 0.5 mm thickness and 6 mm diameter with top and bottom electrodes made out of Silver paste following the annealing process at 500 °C in air for proper adhesion. SGO films with the thickness of ~ 75 nm were grown on $Pt/TiO_2/SiO_2/Si$ substrates at the fixed temperature of 300 °C and under the oxygen partial pressure of 30 mTorr. The KrF excimer laser (248 nm, 10 Hz) operating at a fluence of ~ 2 J/cm^2 was used for the ablation of the ceramic SGO target. The MIM capacitor structures Pt/SGO/Pt were designed after the deposition of Pt-top electrodes with the diameter of 80 μm, through a shadow mask at room temperature by dc-magnetron sputtering. The surface morphology of the films was investigated by atomic force microscopy (AFM). The dielectric measurements were done with HP4294A Impedance Analyzer. I-V characteristics of bulk sample and the switching characteristics of Pt/SGO/Pt thin film devices were analysed using a Keithley 2401 source-meter unit in the top-bottom configuration.

DISCUSSIONS

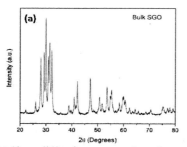 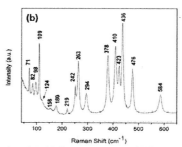

Fig.1 (a) X-ray diffraction pattern of powder ceramic SGO (b) Raman spectra of SGO ceramic at room temperature.

In order to investigate the structural properties we have done the X-ray diffraction analysis and Raman spectroscopic measurements on bulk powder samples as well as on thin films of SGO. The x-ray diffraction pattern of the powder sample has been shown in Fig.1 (a). The diffraction peaks were well indexed with a B-type monoclininc structure (space group *C2/m*) by comparing the x-ray diffraction pattern with the as reported x-ray results of one of the ternary oxide $LaGdO_3$

[6]. As shown in the Fig.1 (b), the powder Raman spectra clearly illustrated the 17 Raman signature modes which confirmed the monoclinic phase at ambient conditions [6]. The SGO thin film x-ray diffraction pattern was shown in Fig.2 (a). Absence of any characteristic peaks clearly demonstrated the amorphous nature of as grown thin films. Fig.2 (b) showed the surface morphology of SGO thin film of the thickness of ~ 75 nm carried out using AFM. The planer and 3-D AFM views of the film demonstrated the smooth and particulates free surface with the root mean square (rms) roughness of 1.6 nm.

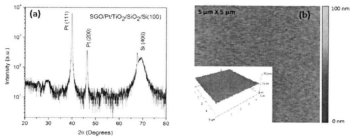

Fig.2 (a) X-ray diffraction pattern of SGO thin film of the thickness of ~ 75 nm deposited at 300°C on Pt/TiO$_2$/SiO$_2$/Si substrate. (b) AFM planar and 3-D view images of SGO film.

Fig.3 (a) depicted the variation in the real part of the dielectric permittivity (ε') and loss tangent (tanδ) as a function of frequency in the range of 100 Hz to 1 MHz at room temperature and values of dielectric constant and the loss has been obtained to be ~ 19 and 0.002, respectively. Fig.3 (b) showed the C-V characteristic of bulk MIM stack at constant frequency of 100 KHz. The negligible variation in capacitance with applied voltage and absence of hysteresis again indicated the linear dielectric behaviour of bulk SGO. The semilog plot of I-V characteristic of bulk MIM stack at room temperature was illustrated in Fig.4 (a). The hysteresis of I-V curve obtained by sweeping the applied bias voltage (the measuring loop followed 0 → 20 → 0 → -20

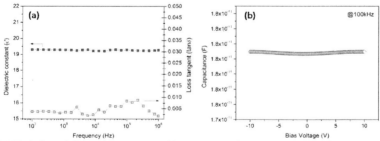

Fig.3 (a) Variation of dielectric constant and loss tangent of SGO ceramic with frequency (100 Hz - 1 MHz range) at room temperature. (b) C-V characteristics of ceramic MIM stack at 100 kHz.

→ 0 cycle), revealed the occurrence of different resistance states during the voltage sweep. The shift of the current minimum towards positive from zero voltage could be interpreted in terms of the effect of internal field due to non-uniform distribution of space charges [7].

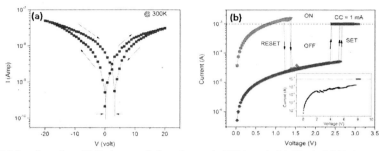

Fig.4 (a) Semilog plot of I-V characteristics of ceramic MIM stack (b) Typical I-V characteristic of the Pt/SGO/Pt thin film ReRAM device showing forming process and resistive switching for three consecutive cycles.

This hysteretic I-V characteristic could be indicated the resistive switching effect in bulk MIM SGO structure. However, we found less abrupt changes in resistance states in our results, but this kind of hysteretic I-V behaviour is found very scarce in bulk oxides, and needs further study. The observed electrical and dielectric results on the bulk samples motivated us to further explore the electrical studies on thin films of this novel high-k dielectric oxide.

We investigated the I-V characteristics of the as fabricated Pt/SGO/Pt thin film device and found reproducible unipolar resistance switching (RS) behaviour with large contrast in ON and OFF resistance states. Fig.4 (b) showed the semi-logarithmic plot of I-V characteristics of as fabricated device. Initially the device stays in an insulating state with resistance of 15 MΩ (read at 0.1 V) until the applied voltage reaches to 8 Volt. At this voltage a sudden increase in current has been observed and the device resistance dropped to a low value of 10 kΩ (read at 0.1 V). This process is called the initial forming process, as shown in the inset of Fig.4 (b). The forming process is necessary to obtain the stable RS behaviour. After the forming process device switched to low resistance state (LRS). In initial forming process current compliance was kept fixed at 1 mA to avoid the dielectric breakdown of the device due to high current flow in LRS. Now the voltage was again swept from 0 to 1.5 V in small steps while measuring the current. We observed the sudden drop in current at a voltage of ~ 1.1 V which indicated the abrupt increase in the resistance of device. This sudden increment in current switched the device from LRS into high resistance state (HRS) as illustrated in Fig.4 (b). This is called the 'reset' process in which the device switched again from ON to OFF-state. The HRS and LRS of the device remained preserved even when the applied bias voltage was removed. In such process, a non-volatile unipolar resistive memory device is demonstrated. Again as the voltage was swept from 0 to 3 V in the HRS of the device, a sudden drastic increase in current was observed at ~ 2.4 V, which switched the device again into LRS by keeping the fixed compliance current of 1 mA. This is called as the 'set' process which switched the device from OFF to ON-state.

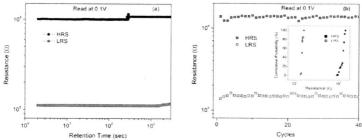

Fig.5 (a) The endurance of the Pt/SGO/Pt ReRAM device over 40 cycles. (b) The retention characteristic of HRS and LRS of the device.

Moreover, form the set/reset cycling stress test over 40 cycles it could be observed that the reset voltage was spread over a small range of voltages between ~ 1.1 and 1.4 V while the set voltage had a spread between ~ 2·2 and 2·7 V. This non-overlapping window between the LRS and HRS values of reset and set voltages clearly demonstrated the utility of our device towards nonvolatile ReRAM applicationsMoreover, the retention characteristics of different resistance states were depicted in Fig.5 (a), indicating that each memory state was preserved stably without degradation over 10^3 Seconds. The endurance of the device under fixed compliance current of 1 mA has been performed up to 40 cycles, as shown in Fig.5 (b), where the resistances of the ON and OFF states were measured at 0.1 V. The statistical distribution of these two resistance states has been depicted in the inset of Fig.5 (b).

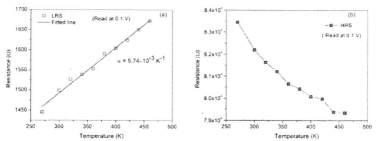

Fig.6 (a) Temperature dependence of resistance in (a) ON-state (LRS) and (b) OFF-state (HRS) of the device.

We measured the temperature dependent resistance values of HRS state (R_{OFF}) and LRS state (R_{ON}) to discus about the switching mechanism responsible for the RS behaviour. Fig.6 (a) illustrated the variation of R_{ON} (read at 0.1 V) with respect to temperature in the range of 270 – 460 K, whereas the temperature variation of R_{OFF} has been shown in the Fig.6 (b). The R_{ON} was found to increase with temperature which is typical for electronic transport in metal, and therefore elucidate the formation of metallic filament after the set operation. The variation of this metallic resistance with temperature can be formulated as; $R_T = R_O[1 + \alpha(T - T_O)]$, where, R_O is the resistance at T_O and α is the resistance temperature coefficient. The above relation is used to fit the variation in R_{ON}, and thus the calculated values of resistance temperature coefficient has been obtained as; $\alpha = 5.74 \times 10^{-3} K^{-1}$. The order of α is in agreement with the resistance temperature coefficients of metallic nanowires. This behaviour of R_{ON} with temperature indicated that the metallic behaviour of the ON-state originates from the formation of conducting filaments, whereas the resistance in OFF-state of the device decreases with temperature which exhibited a semiconducting behaviour, as shown in Fig.6 (b). On the basis of the above temperature dependent electrical results, it could be concluded that the switching mechanism in the device followed the conductive filament model.

CONCLUSIONS

We studied the structural and electrical characteristics of the novel ternary high-k dielectric oxide SGO in its bulk and thin film forms. The B-type monoclinic structure of ceramic SGO has been confirmed by X-ray diffraction and Raman spectroscopic measurements, whereas the pulsed laser deposited thin films were found to be amorphous. The electrical and dielectric measurements on bulk samples revealed a linear dielectric constant of ~ 19 with low dielectric loss and hysteretic leakage current. In addition, we proposed the ReRAM device with Pt/SGO/Pt thin film capacitor structure and studied the unipolar-RS behaviour. The switching mechanism for the RS behaviour was explained in the light of conductive filament model. The observed nonvolatile resistive switching with sufficient ON/OFF resistance ratio, good endurance as well as retention, high dielectric constant, and amorphous structure make SGO as a promising material for logic and memory applications.

ACKNOWLEDGEMENT

The authors acknowledge financial support from DOE (grant DE-FG02-ER46526).

REFERENCES

1. I. Warshaw, and R. Roy, J. Phys. Chem. 65, 2048 (1961).
2. W.C. Chin, K. Y. Cheong, Z. Hassan, Mat. Sc. in semiconductor processing 13, 303 (2010).
3. L. Chen, W. Yang, Y. Li, Q. Sun, P. Zhou, H. Lu, S. Ding, and D. W. Zhang, J. Vac. Sci. Technol. A 30, 01A148 (2012).
4. Y. Zhao, Materials 5, 1413 (2012).
5. S. P. Pavunny, P. Misra, R. Thomas, A. Kumar, J. Schubert, J. F. Scott, and R. S. Katiyar, Appl. Phys. Lett. 102, 192904 (2013).
6. S. P. Pavunny, A. Kumar, P. Misra, J. F. Scott, and R. S. Katiyar, Phys. Status Solidi B, 1–9 (2013).
7. Y. Xia, W. He, L. Chen, X. Meng, and Z. Liu, Appl. Phys. Lett. 90, 022907 (2007).

Correlation of resistance switching behaviors with dielectric functions of manganite films: A study by spectroscopic ellipsometry

Masaki Yamada[1], Toshihiro Nakamura[2], and Osamu Sakai[1]
[1] Department of Electronic Science and Engineering, Kyoto University, Kyotodaigaku-Katsura, Nishikyo-ku, Kyoto 615-8510, Japan
[2] Department of Engineering Science, Osaka Electro-Communication University, 18-8 Hatsu-cho, Neyagawa, Osaka 572-8530, Japan

ABSTRACT

$Pr_{0.5}Ca_{0.5}MnO_3$ (PCMO) films were deposited on $LaAlO_3$ (100) substrates under pressure from 1.33 to 5.33 Pa by RF magnetron sputtering. Resistance switching and dielectric functions of PCMO films were studied by DC current-voltage characteristic measurements and spectroscopic ellipsometry (SE) measurements. Resistance switching was observed in the devices composed of PCMO films deposited under low pressures of 1.33 and 2.67 Pa. SE measurements revealed that dielectric functions also depended on deposition pressure. PCMO films deposited under lower pressure had larger high-frequency dielectric constant, larger oscillator strength of the electric dipole charge transitions in MnO_6 octahedral complexes, and lower oscillator strength of d-d transitions in Mn^{3+} and Mn^{4+} ions. SE measurements suggested that oxygen vacancies and MnO_6 octahedral complexes play an important role in resistance switching in PCMO films.

INTRODUCTION

Recently, next-generation nonvolatile memories have been intensively researched and developed as future replacements for volatile memories such as dynamic random access memory in the application to IT equipment. Electric-pulse-induced resistance switching observed in metal oxides such as $Pr_{1-x}Ca_xMnO_3$ provides a possibility of one of next-generation nonvolatile memories, called resistance random access memory (ReRAM) [1-8]. ReRAM has the advantage of low power consumption, small bit cell size, and fast switching speed. The ReRAM devices composed of $Pr_{1-x}Ca_xMnO_3$ films show bipolar resistance switching behavior. Impedance spectroscopic studies suggested that resistance switching was mainly due to the resistance change in the interface between the $Pr_{1-x}Ca_xMnO_3$ films and the electrode [4,5]. Moreover, the transport of oxygen ions between the interfacial oxide layer of electrode materials and oxygen vacancies by the electric field was proposed as a model of the resistance switching [6,7]. On the other hands, resistance switching behaviors are affected by the crystal structure of $Pr_{1-x}Ca_xMnO_3$ films [8]. The crystal structure of $Pr_{1-x}Ca_xMnO_3$ films might affect the behavior of oxygen ions migrating between the $Pr_{1-x}Ca_xMnO_3$ films and the interfacial oxide layer. Considering that the crystal structure has a significant influence on optical properties in films, spectroscopic ellipsometry (SE) is one of the useful methods for studying optical properties of various materials including perovskite manganese oxide [9,10]. In SE, the optical properties of thin films can be analyzed on the basis of dielectric functions. The dielectric function in the visible light region gives much information on microstructure and electronic band structure [11,12].

In this work, $Pr_{0.5}Ca_{0.5}MnO_3$ (PCMO) films were deposited under pressure from 1.33 to 5.33 Pa by RF magnetron sputtering. The deposition pressure dependence of the resistance switching behavior and electric properties in PCMO-based devices was investigated by current-voltage (I-V) measurements. The optical properties of PCMO films deposited under various pressures were analyzed as the dielectric functions by SE. The correlation of resistance switching behavior with the optical properties was discussed on the basis of the I-V characteristics and dielectric functions of the PCMO films.

EXPERIMENTAL DETAILS

PCMO films were deposited on $LaAlO_3$ (100) substrates by RF magnetron sputtering. A mixture of 75 % Ar and 25 % O_2 gases was used as the sputtering gas. The deposition pressures were 1.33, 2.67, 4.00, and 5.33 Pa. The input RF power was 80 W. The substrate temperature was 650 °C. The film thickness and surface roughness of the films were measured by atomic force microscopy (AFM). AFM measurements showed that the film thickness was increased from 38 to 57 nm as the deposition pressure was increased. The averaged roughness was ranged from 0.1 to 0.5 nm depending on deposition pressure.

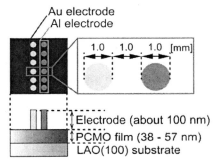

Figure 1. Schematic diagram of PCMO-based resistance switching devices.

In order to measure the electric properties of the deposited films, aluminium (Al) and gold (Au) electrodes were deposited on top of PCMO films by vacuum evaporation. The intervals between two electrodes were 1.0 mm, and the diameters of the electrodes were 1.0 mm. The thicknesses of the electrodes were about 100 nm. Figure 1 shows the schematic diagram of PCMO-based devices. In I-V measurements, the positive voltage is defined as the current flows from Al electrode to Au electrode via PCMO film, and negative one was defined by the opposite direction.

The ellipsometric spectra were measured by spectroscopic ellipsometer (FE-5000, Otsuka Electronics) at the photon energy region from 1.75 to 4.00 eV. The dielectric function of PCMO films was determined by comparing experimental spectra with the simulated ones based on triple Lorentz oscillators model

$$\varepsilon = \varepsilon_\infty + \sum_{j=1}^{3} \frac{f_j \omega_{0j}^2}{\omega_{0j}^2 - \omega^2 + i\gamma_j \omega}$$

where ε_∞ was the high-frequency dielectric constant, f_j the oscillator strength parameter, ω_{0j} the oscillator frequency, γ_j the damping factor.

DISCUSSION

I-V characteristics of PCMO-based devices

The *I-V* characteristics of the PCMO-based devices were studied by DC voltage sweep measurements to study the deposition pressure dependence of resistance switching behavior. Figure 2 shows the *I-V* characteristics of the PCMO-based devices. The voltage bias was swept as $0 \rightarrow +10 \rightarrow 0 \rightarrow -10 \rightarrow 0$ V. When the positive bias was applied, the resistance of the devices composed PCMO films deposited under low pressure of 1.33 Pa and 2.67 Pa was increased in 2.0 – 4.0 V. This behavior corresponded to the resistance switching from low resistance state (LRS) to high resistance state (HRS). When the negative bias was applied, the

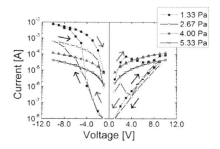

Figure 2. *I-V* characteristics of PCMO-based devices. The PCMO films were deposited under pressure from 1.33 to 5.33 Pa.

resistance switching from HRS to LRS occurred in the devices composed PCMO films deposited under low pressure of 1.33 Pa and 2.67 Pa. On the other hands, the devices composed PCMO films deposited under high pressure of 4.00 Pa and 5.33 Pa showed no hysteresis loop, resulting non-switching behavior.

Optical properties of PCMO films

Figure 3 shows the dielectric functions of PCMO films obtained from ellipsometric spectra of the PCMO films. Different dielectric functions were obtained depending on the deposition pressure. The determined parameters of the triple Lorentz oscillators model for dielectric functions are listed in Table 1. The oscillator frequencies of OS1, OS2, and OS3 were 3.88 – 4.09, 2.52 – 2.96, and 0.55 – 0.82 eV in unit for energy, respectively. The PCMO films deposited at low pressures of 1.33 and 2.67 Pa showed large oscillator strength of OS1, small oscillator strength of OS2, and high ε_∞ value. As the deposition pressure was increased, the oscillator strength of OS1 was decreased, the oscillator strength of OS2 was increased, and the ε_∞ value was decreased. We suggested that the deposition pressure dependence of electronic state of PCMO films was detected as a difference in the parameters of the dielectric functions.

For ABO$_3$ perovskites, the top of the valance band (VB) is mainly made up from O $2p$, and the bottom of the conduction band (CB) consists of the $d\ t_{2g}$ of the B-site atoms [13]. The p-d hybridization orbital between O and B-site atoms is at the maximum of VB [14]. The interband electronic transition at the lowest band gap is suggested that the electric dipole transition in BO$_6$ octahedral structure. Transition metal ions at B-site of perovskites have the ligand field transition from $d\ t_{2g}$ to $d\ e_g$. La$_{1-x}$Sr$_x$MnO$_3$ (LSMO) has d-d ligand field transitions in the Mn^{3+} (2.5 and 2.6 eV) and Mn^{4+} (3.1 eV) and electric dipole charge transitions from O $2p$ to Mn $3d$ in the (MnO$_6$)$^{9-}$ (3.5 eV) and (MnO$_6$)$^{8-}$ (4.3 eV) octahedral complexes [12]. As for the d-d ligand field transitions, the signal at 2.5 eV was assigned to the $^5E_g - {}^5T_{2g}$ transition in Mn^{3+}, while the signals at 2.6 eV and 3.1 eV were assigned to the $^4A_{2g} - {}^4T_{2g}$ and $^4A_{2g} - {}^4T_{1g}$ transitions in Mn^{4+}, respectively [15].

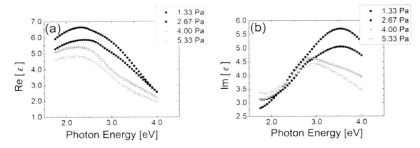

Figure 3. (a) Real (Re[ε]) and (b) imaginary (Im[ε]) parts of dielectric functions of PCMO films. The PCMO films were deposited under pressure from 1.33 to 5.33 Pa.

Table 1. Parameters of the triple Lorentz oscillators model for PCMO films deposited under pressure from 1.33 to 5.33 Pa. The film thickness data determined by SE are compared with the values measured by AFM.

deposition pressure [Pa]		1.33	2.67	4.00	5.33
ε_∞		3.42	3.13	2.58	2.55
OS1	f_1	4.42	4.14	4.06	3.85
	$\hbar\omega_{01}$ [eV]	3.88	3.98	4.09	4.04
	$\hbar\gamma_1$ [eV]	3.18	3.53	4.52	4.91
OS2	f_2	0.05	0.04	0.28	0.37
	$\hbar\omega_{02}$ [eV]	2.52	2.75	2.77	2.96
	$\hbar\gamma_2$ [eV]	0.58	0.48	0.90	1.31
OS3	f_3	21.27	23.11	10.99	8.32
	$\hbar\omega_{03}$ [eV]	0.55	0.56	0.69	0.82
	$\hbar\gamma_3$ [eV]	0.48	0.84	0.79	0.78
Film thickness d [nm] (SE)		55	50	48	37
Film thickness d [nm] (AFM)		57	54	44	38

In reference to the electronic structure of LSMO, the transitions corresponding to OS1 and OS2 of the PCMO films were attributed to the electric dipole charge transitions from O 2p to Mn 3d in $(MnO_6)^{9-}$ and $(MnO_6)^{8-}$ octahedral complexes and the ligand field d-d transition in Mn^{3+} and Mn^{4+} ions, respectively. Figure 4 shows the schematic diagram of electronic band structures and the interband transition corresponding to OS1. The PCMO films deposited under lower pressure had larger oscillator strength of the electric dipole charge transition in the $(MnO_6)^{9-}$ and $(MnO_6)^{8-}$ octahedral complexes, and smaller oscillator strength of d-d transitions in

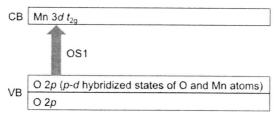

Figure 4. Electronic band structures and interband transition from O $2p$ to Mn $3d$ in $(MnO_6)^{9-}$ and $(MnO_6)^{8-}$ octahedral complexes. This transition corresponds to OS1.

the Mn^{3+} and Mn^{4+} ions. As the deposition pressure is decreased, sputtered species arrive on depositing film with higher energy and can form $(MnO_6)^{9-}$ and $(MnO_6)^{8-}$ octahedral complexes, and the imaginary part of ε ($Im[\varepsilon]$) in the electric dipole charge transition in the $(MnO_6)^{9-}$ and $(MnO_6)^{8-}$ octahedral complexes was increased. As the $(MnO_6)^{9-}$ and $(MnO_6)^{8-}$ octahedral complexes were formed in the films, the oscillator strength of the electric dipole charge transition in the $(MnO_6)^{9-}$ and $(MnO_6)^{8-}$ octahedral complexes was increased, and the oscillator strength of d-d transitions in the Mn^{3+} and Mn^{4+} ions was decreased because manganese ions were reduced to Mn^{2+} ions, whose d-d transitions are forbidden, by removing oxygen ions in the structures of $(MnO_6)^{9-}$ and $(MnO_6)^{8-}$ octahedral complexes. These result suggested that PCMO films deposited lower pressure had higher density of $(MnO_6)^{9-}$ and $(MnO_6)^{8-}$ octahedral complexes. Considering that resistance switching is also observed in the devices composed PCMO films deposited under lower pressure, the formation of $(MnO_6)^{9-}$ and $(MnO_6)^{8-}$ octahedral complexes is important for obtaining large resistance switching.

On the other hand, high-frequency dielectric constant ε_∞ indicated the metallicity of an oxide. Large high-frequency dielectric constant in the PCMO films deposited at lower pressure corresponded to larger conductivity. I-V measurements showed that the resistance of the deposited films was decreased, as the deposition pressure was decreased. Since oxygen vacancies work as donors in oxide [16], PCMO films deposited under lower pressure had higher density of oxygen vacancies. Considering that PCMO films deposited under lower pressure have larger resistance change and larger high-frequency dielectric constants, oxygen vacancies are also required for large resistance switching.

CONCLUSIONS

PCMO films were deposited on $LaAlO_3$ (100) substrates under pressure from 1.33 to 5.33 Pa by RF magnetron sputtering. Resistance switching was observed in the devices composed of the PCMO films deposited at low pressures of 1.33 and 2.67 Pa. The deposition pressure dependence of the electronic structure of PCMO films was detected as a difference in dielectric functions by SE. SE data indicated that the PCMO films exhibiting resistance switching had large oscillator strength of the electric dipole charge transition in the $(MnO_6)^{9-}$ and $(MnO_6)^{8-}$ octahedral complexes, small oscillator strength of d-d transitions in the Mn^{3+} and Mn^{4+} ions, and large high-frequency dielectric constant. The $(MnO_6)^{9-}$ and $(MnO_6)^{8-}$ octahedral complexes are suggested to work as a supplier and receiver of oxygen ions and enable oxygen ions to move between the PCMO film and the interfacial oxide layer of the electrode materials. Oxygen vacancies are suggested to help oxygen ions moving by electric fields. SE is a promising tool for non-destructive predicting whether a device shows resistance switching or not.

ACKNOWLEDGEMENTS

This work was supported in part by Grant-in-Aid for Challenging Exploratory Research (No. 23656215) from the Japan Society for the Promotion of Science (JSPS). A part of this work was supported by Kyoto University Nano Technology Hub in "Nanotechnology Platform Project" sponsored by the Ministry of Education, Culture, Sports, Science and Technology (MEXT), Japan.

REFERENCES

1. S. Q. Liu, N. J. Wu, A. Ignatiev, *Appl. Phys. Lett.* 76 (2000) 2749.
2. A. Baikalov, Y. Q. Wang, B. Shen, B. Lorenz, S. Tsui, Y. Y. Sun, Y. Y. Xue, C. W. Chu, *Appl. Phys. Lett.* 83 (2003) 957
3. A. Sawa, T. Fujii, M. Kawasaki, Y. Tokura, *Appl. Phys. Lett.* 85 (2004) 4073.
4. T. Nakamura, K. Homma, K. Tachibana, *J. Nanosci. Nanotechnol.* 11 (2011) 8408.
5. T. Nakamura, K. Homma, K. Tachibana, *Nanoscale Res. Lett.* 8 (2013) 76.
6. A. Sawa, *Mater. Today*, 11 (2008)
7. M. Siddik, K. P. Biju, X. Liu, J. Lee, I. Kim, S. Kim, W. Lee, S. Jung, D. Lee, S. Sadaf, H. Hwang, *Jpn. J. Appl. Phys.*, 50 (2011) 105802.
8. H. Yamada, A. Sawa, *Phys. Rev. B*, 80 (2009) 235113.
9. L. V. Nomerovannaya, A. A. Makhnëv, and A. Yu. Rumyantsev, *Phys. Solid State*, 41 (1999) 1322.
10. N. N. Loshkareva, L. V. Nomerovannaya, E. V. Mostovshchikova, A. A. Makhnev, Yu. P. Sukhorukov, N. I. Solin, T. I. Arbuzova, S. V. Naumov, and N. V. Kostromitina, A. M. Balbashov and L. N. Rybina, *Phys. Rev. B* 70 (2004) 224406.
11. E. D. Gaspera, S. Schutzmann, M. Guglielmi, A. Martucci, *Opt. Mater.*, 34 (2011) 79-84.
12. N. N. Loshkareva, Yu. P. Sukhorukov, E. V. Mostovshchikova, L. V. Nomerovannaya, A. A. Makhnev, S. V. Naumov, E. A. Gan'shina, I. K. Rodin, A. S. Moskvin, A. M. Balbashov, *J. Exp. Theor. Phys.*, 2 (2002) 350.
13. W. J. Zhang, Z. H. Duan, K. Jiang, Z. G. Hu, G. S. Wang, X. L. Dong, J. H. Chu, *Acta Mater.*, 60 (2012) 6175.
14. H. W. Eng, P. W. Barnes, B. M. Auer, P. M. Woodward, *J. Solid State Chem.*, 175 (2003) 94.
15. E.A. Balykina, E.A. Ganshina, G.S. Krinchik, A.Yu. Trifonov, I.O. Troyanchuk, *J. Magn. Magn. Mater.*, 117 (1992) 259.
16. N. Shanthi, D. D. Sarma, *Phys. Rev. B*, 57 (1998) 2153.

A continuous composition spread approach towards monolithic, wavelength-selective multichannel UV-photo-detector arrays

H. von Wenckstern, Z. Zhang, J. Lenzner, F. Schmidt and M. Grundmann
Universität Leipzig, Fakultät für Physik und Geowissenschaften, Institut für Experimentelle Physik II, Linnéstrasse 5, 04103 Leipzig, Germany

ABSTRACT

Continuous composition spread (CCS) methods have been very successfully used for exploiting and optimization of new material systems. Concerning sample growth by pulsed-laser deposition (PLD) approaches towards thin films with a CCS are involved, here movable masks for partial shadowing of the substrate and multiple targets are needed to obtain linearly varying changes of composition. Here we make use of an approach allowing deposition of thin films with CCS at high growth rates by using segmented PLD targets. We describe how this approach can be used to fabricate monolithic, wavelength-selective multichannel UV-photo-detector arrays.

INTRODUCTION

The detection of ultraviolet irradiation is important in environmental research and monitoring, flame detection and monitoring of industrial processes as UV curing of glues, adhesives or disinfection of drinking water by UV irradiation. Besides (Al,Ga)N semiconducting oxides present a material class that is very suited for realization of detectors operating in the UV-A, UV-B and even the UV-C spectral range. For most applications the determination of spectrally integrated UV radiation is not sufficient and a spectrally resolved detection of UV radiation is desired. Here, we demonstrate an approach towards monolithic wavelength-selective UV-A photo-detectors (PD) by using a continuous composition spread (CCS) approach.

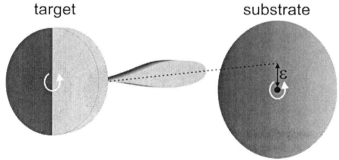

Fig. 1: Schematics of the single target CCS-PLD method. Here, the case of a two-fold segmented target is depicted. Target and substrate rotation are synchronized. The center of the plasma

plume must have a spatial offset ⍵ with respect to the center of the substrate for CCS formation.

Thin films with CCS have been created by various physical deposition methods such as magnetron sputtering and pulsed-laser deposition. Recently, we introduced a facile CCS technique for pulsed-laser deposition (PLD) that makes use of segmented PLD targets[1]. The use of such targets (cf. **Fig. 1**) makes partial shadowing of the substrates and time-consuming target changes as necessary in conventional PLD CCS approaches[2] obsolete. In our segmented target CCS-PLD technique the growth rate, thickness distribution and the influence of the growth temperature is the same as for the deposition of laterally homogeneous thin films from a standard PLD target. Therefore, experience that has been gathered over long periods of time and that has been used to optimize an existing PLD setup is not lost but can instead be used to optimize thin film growth from segmented PLD targets. The resulting compositional gradient on the substrates depends on the background pressure (on the kinetics of the plasma expansion) and on geometrical parameters in particular the target-to-substrate distance z and the offset ⍵

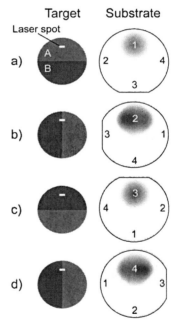

Fig. 2: Illustration of the creation of a one-dimensional composition spread using a two-fold segmented target. a) Ablation of constituent A and schematic representation of the distribution of elements from target component A on substrate position 1. b) The target and substrate are now rotated by 90° with respect to a). Now both target segments are ablated; constituents of segment A and B are stronger distributed towards substrate positions 1 and 3, respectively. c) Target and substrate are rotated by another 90°. Constituents of segment B are primarily incorporated at substrate location 3. d) After another rotation by 90° again both segments are ablated and again constituents of segment A and B are stronger distributed towards substrate positions 1 and 3, respectively.

between the centerline of the expanding plasma plume and the rotational axis (center) of the substrate. In addition, there are conditions for obtaining homogeneous film thickness across larger substrates that have to be considered. Typically, substrate rotation and an off-center deposition (meaning ⌀⌀> 0) reduce thickness variations to an acceptable value (better that 90%). Further, it is desirable to rotate the target during deposition as well. This assures that the target is effectively used up and that the target does not undergo strong local heating. For CCS-PLD the target and the substrate must be rotated synchronized and ⌀⌀> 0 is actually a necessary condition[1].

EXPERIMENT

In a PLD process the laser spot on the target is small compared to the target's surface area. Hence, the composition of the plasma plume varies in time. Now, we want to illustrate with help of Fig. 2 how this periodically changing particle flux results in the creation of a lateral composition gradient for a two-fold segmented target. For that, we investigate the composition of the plasma plume and the composition of impinging particle flux on the substrate for four

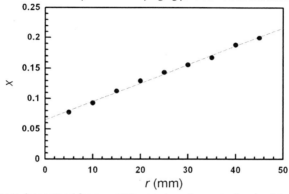

Fig. 3: Mg-content determined from an EDX scan along the centerline (and the compositional gradient) of an $Mg_xZn_{1-x}O$ CCS-PLD thin film. The dashed line is a linear fit to the experimental data.

different angular positions of target and substrate. If the laser spot is at one time incident to segment A (cf. Fig. 2(a)), the plasma plume has a composition similar to that of segment A and the highest flux of particles of this composition onto the substrate occurs (opposite to the incident laser spot) on substrate location 1. As both the target and the substrate rotate the substrate location encountering the highest particle flux changes. After a rotation of 90° (cf. Fig. 2(b)) the laser spot is incident to both target segments. The composition of the plasma plume changed and now it is a mixture of elements from segment A and segment B. Therefore, elements from both segments are incident on substrate position 2, however, particles from segment A are stronger distributed towards the right side of the substrate (towards position 1) and elements originating from segment B are more distributed towards substrate position 3. After half a rotation of target and substrate the composition of the plasma plume corresponds to that of segment B. Due to the rotation the substrate location now encountering the highest flux is exactly opposite (or rotated by 180°) to the location that encountered the highest flux of particles with composition A (cf. Fig. 2c). After another rotation by 90° again both segments are ablated as depicted in Fig. 2d). This time constituents if segment B are stronger distributed to the right side of the substrate which corresponds now to substrate position 3. Correspondingly, elements of segment A are now more likely incident on the left side of the substrate (substrate position 1). In short one can state that because of the synchronized target and substrate rotation and because of the fact that the center of the plasma plume is not incident to the center of the substrate the deposited thin film will have a lateral composition spread.

Based on this facile approach to create thin films with defined lateral change of the alloy composition photo-detector arrays for which the onset of absorption depends on the lateral position within the array are feasible. In a first attempt we used targets with two segments that consisted of segment A: $Mg_{0.02}Zn_{0.98}O$ and segment B: $Mg_{0.1}Zn_{0.9}O$. The lowest (highest) Mg-content in the resulting sample is $x \approx 0.06$ ($x \approx 0.21$). Along the centerline, which is in the direction of the gradient, the composition changes linearly across the 2-ich substrate for the process parameters used (target-to-substrate distance 90 mm, background oxygen pressure 0.016 mbar, ▨ ≈ 20 mm). In Fig. 4(a) we depict the normalized responsivity of selected Pd/ZnO/Pd-photo-detectors[3,4] prepared on different positions of the 2-inch (Mg,Zn)O thin film. As expected from the EDX data of Fig. 3, the onset of the absorption E_{cutoff} depends on the lateral position. Fig. 3(b) is a false color representation of the dependence of E_{cutoff} on the lateral position within that 2-inch wafer. The letters a-e represent the position of the photo-detectors whose photo response is depicted in Fig. 4(a). From Fig. 4(b) we find that the band-gap (and with that the Mg-content) changes systematically in a direction parallel to the gradient while it is in principle constant along directions perpendicular to the gradient. This first example illustrates first and foremost that it is easily possible to vary the absorption edge and with that the onset of the photo-response of (Mg,Zn)O PDs linearly across a 2-inch in diameter thin film. Especially the linearity within the composition spread makes this approach interesting for more involved photo-detector designs.

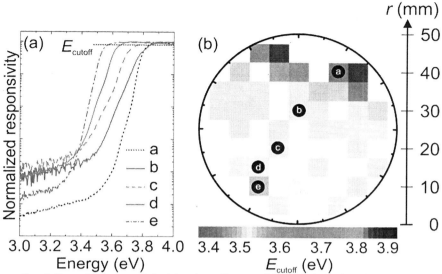

Fig. 4: Normalized responsivity (a) and cutoff energy (b) of MSM-UV-photodetectors on (Mg,Zn)O CCS-PLD thin film. The lateral change of composition causes the spatial dependence of absorption edge and with that the onset of photoresponse (E_{cutoff}).

In a next step, one can include an optical filter in the design of the photo-detectors in order to limit the detectivity to a narrow energy range, which is defined by the difference of the absorption edge of the filter and that of the active layer. In a straightforward approach the active layer is deposited directly onto the filter layer. If the filter layer has a constant Mg-content, the action of the active CCS-PLD layer would result in a spatial dependence of the onset of the photo-response. The onset of the absorption in the filter layer would not have a lateral dependence and hence "only" the bandwidth of such a band-pass detector would depend on the position[5]. A more elaborate approach would comprise a compositional gradient within both the filter layer and within the active layer. If the gradient in the filter layer and that in the active layer are parallel to each other (and if the band-gap of the filter layer is for each position higher than that in the active layer) wavelength-selective, monolithic photo-detector arrays are obtainable. However, there is an obstacle within this approach that needs to be considered. Typically, the crystal quality of semiconductors depends strongly on their deposition temperature and for wide band-gap semiconductors based on (Mg,Zn)O the growth temperature typically exceeds 500°C. Here, interdiffusion will occur resulting in a softening of the absorption edge of the active as well as the filter layer. To our opinion the most effective way to suppress this effect is to deposit the filter layer on the backside of the substrate leading to a spatial decoupling of the active and the filter layer. This is schematically illustrated in Fig. 5 in which the bandwidth of PDs having the same Mg-content in the active and the same in the filter layer is depicted. Since interdiffusion is turned off if the filter and the active layer are deposited on opposite sides of the substrates the bandwidth of such PDs will always be lower (for given Mg-contents) than that of PDs having filter and active layer on top of each other. Therefore, it is recommended make use of the deposition on opposite substrate sides if the energetic resolution of a PD-array is crucial.

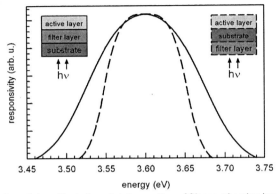

Fig. 5: Illustration of the effect of spatial decoupling of filter and active layer. In the latter approach the filter is deposited on one side of the substrate, then the substrate is flipped and then the active layer is deposited. In result, interdiffusion between the active and the filter layer is suppressed and the minimal bandwidth of the wavelength-selective photo-detectors decreases and hence energetic resolution of a PD-array improves.

With that, all ingredients necessary for the realization of monolithic, wavelength-selective multi-channel UV-photo-detector arrays are on the table and only have to be combined. We suggest the following recipe for narrow bandwidth PDs: i) deposition of the filter layer at background pressure p_1 from a segmented PLD target with rather small difference in the Mg-content (e.g. segment A: $Mg_{0.02}Zn_{0.98}O$ and segment B: $Mg_{0.1}Zn_{0.9}O$) and ii) deposition of the active layer at background pressure $p_2 > p_1$ from the same segmented PLD target. Due to the increased background pressure used for the deposition of the active layer, the Mg-content in the active layer will be slightly lower (for higher background pressures the lighter Mg atoms are more effectively scattered during the expansion of the plasma plume and hence a lower Mg flux is incident to the substrate).

Finally we want to comment on the steepness of the two flanks of the narrow-bandwidth PDs. The area of PDs fabricated in our laboratory is about 300 × 300 µm². For the thin film of Fig. 3 the lateral composition spread results in a change of the Mg-composition within 300 µm of roughly 0.001 which means that the band-gap within the active layer of one detector changes by almost 2 meV. The same holds true for the optical filter layer. That is why we suggested rather moderate gradients for the design of monolithic, wavelength-selective multichannel PD-arrays. Further this explains why it is impossible to go beyond a certain steepness of the absorption edge (cf. Fig. 5) for the active as well as for the filter layer.

For commercial exploitation of such wavelength-selective UV-PDs dimensions of (10 µm)^2 are typical. This implies that i) much stronger compositional variations are feasible without loss of resolution while ii) the spectral response of the detector should be even sharper due to negligible changes of the alloy composition within a single PD.

CONCLUSIONS

In summary we have described a route for the realization of monolithic, wavelength-selective photo-detector arrays by employing a novel composition spread approach relying on the ablation of segmented PLD targets. A design in which the active and the filter layer are deposited on opposite sides of the substrates is preferred becomes it circumvents interdiffusion between the active and the filter layer that would lead to a broadening of the response of the detector. We have discussed limitations concerning the minimal achievable sharpness of the absorption edge of the active layer and the filter layer and showed that these are directly connected to the strength of the compositional gradient within the sample. Our approach is straightforwardly extendable to other materials classes such as $(In,Ga)_2O_3$, $(Al,Ga)_2O_3$ and so on.

ACKNOWLEDGMENTS

The authors thank G. Ramm for target preparation, H. Hochmuth and M. Lorenz for sample growth (all Universität Leipzig). Parts of this work have been supported by Deutsche Forschungsgemeinschaft in the framework of Sonderforschungsbereich 762 "Functionality of Oxide Interfaces".

REFERENCES

[1] H. von Wenckstern, Z. Zhang, F. Schmidt, J. Lenzner, H. Hochmuth and M. Grundmann, CrystEngComm **15**, 10020 (2013)
[2] H. M. Christen, S. D. Silliman, and K. S. Harshavardhan, *Rev. Sci. Instrum.*, 2001, **72**, 2673.
[3] A. Lajn, H. von Wenckstern, Z. Zhang, C. Czekalla, G. Biehne, J. Lenzner, H. Hochmuth, M. Lorenz, M. Grundmann, S. Künzel, C. Vogt, R. Deneke, J. Vac. Sci. Technol. B **27**, 1769 (2009)
[4] Z. Zhang, H. von Wenckstern, M. Schmidt and M. Grundmann, Appl. Phys. Lett. **99**, 083502 (2011)
[5] Z. Zhang, H. von Wenckstern and M. Grundmann, Appl. Phys. Lett. **103**, 171111 (2013)

Metal-Semiconductor-Insulator-Metal Structure Field-Effect Transistors Based on Zinc Oxides and Doped Ferroelectric Thin Films

Ze Jia[1*], Jianlong Xu[2], Xiao Wu[1], Mingming Zhang[2], Naiwen Zhang[2], Jizhi Liu[1], Zhiwei Liu[1], Juin J. Liou[3]

[1] School of Microelectronics and Solid-State Electronics, University of Electronic Science and Technology of China, Chengdu, Sichuan, 610054, China

[2] Institute of Microelectronics, Tsinghua National Laboratory for Information Science and Technology, Tsinghua University, Beijing, 100084, China

[3] Department of Electrical Engineering and Computer Science, University of Central Florida, Orlando, Florida 32816, USA

*Corresponding author's E-mail: ze.jia@ieee.org

ABSTRACT

Different ferroelectric thin films and their related Metal-Semiconductor-Insulator-Metal (MSIM) structures include zinc oxide (ZnO) are studied, which can be utilized in back-gated ferroelectric field-effect transistors (FETs). The most ideal zinc oxide (ZnO) thin film prepared by sol-gel method are obtained under the pyrolysis temperature of 400°C and the annealing temperature of 600°C. The asymmetric or symmetric current-voltage characteristics of the heterostructures with ZnO are exhibited depending on different ferroelectric materials in them. The curves of drain current versus gate voltage for MSIM-structure FETs are investigated, in which obvious counterclockwise loops and a drain current switching ratio up to two orders of magnitude ate observed due to the modulation effect of remnant polarization on the channel resistance. The results also indicate the positive influences of impurity atom substitution in bismuth ferrite thin film for the MSIM-structure FETs.

INTRODUCTION

The back-gated FET is considered as one of the most promising candidates due to its advantages of processing easily for its simple structures, reducing the electrostatic coupling from drain to the channel, controlling threshold voltage dynamically and decreasing static leakage current [1,2]. Among various gate dielectric materials for back-gated FETs, ferroelectric materials have been widely utilized due to their high dielectric constant and their switchable remnant polarization [3,4]. The heterostructures consisting of ferroelectric and semiconductor become the critical part of the back-gated ferroelectric FETs [5-7]. ZnO as a natural n-type semiconductor is introduced as the channel layer because of its relatively high electrical conductivity, low crystallization temperature, good integration with different materials and low costs [8-10]. Among various ferroelectric materials, lead zirconate titanate (PZT) exhibits its medium remanent polarization and suitable crystallization temperature, but its intrinsic toxic lead can limit its further application [11-14]. A lead-free multiferroelectric material, bismuth ferrite

(BFO), has attracted more attentions for application due to its giant remanent polarization at room temperature (RT), high Curie temperature, high Neel temperature, and low crystallization temperature [15,16], though its leakage current density is relatively high [17,18], which can be attributed to the inherent volatility of Bi atoms and valence fluctuation of Fe ions (Fe^{3+} state to Fe^{2+} state) in BFO, generating the oxygen vacancies for charge compensation [19–21]. To suppress the leakage current density of BFO, the impurity atom substitution at the Bi and Fe sites to introduce a chemical pressure into the crystal is considered as one of the most effective methods in optimizing both its crystalline and electronic structures [21]. In this work, different ferroelectric thin films, such as PZT, BFO without dopants, with the dopants of 5% lanthanum (BLFO) and 5% manganese (BFMO) respectively, are studied for their influences on related heterojunctions and MSIM structures, which can be utilized in back-gated ferroelectric FETs shown in Figure 1.

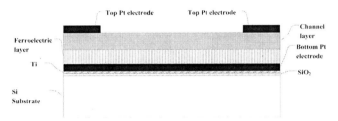

Figure 1. Cross section of back-gated ferroelectric FET based on MSIM structure

EXPERIMENT

Different ZnO thin films were fabricated by sol-gel process on SiO_2 substrate. The $Al(NO_3)_3 \cdot 9H_2O$ was used for doping with the 0.75mol/L ZnO solution which are purchased from Alfa Aesar, A Johnson Matthey Company. These solutions were spin-coated at 1500 rotations per minute (rpm) for 30s and pyrolyzed at 250°C, 300°C, 350°C, 400°C, 450°C for 5min in the air, respectively. The coating and pyrolysis processes were repeated six times, then these thin films were introduced to a rapid thermal annealing (RTA) process at 500°C, 550°C, 600°C, 650°C, 700°C, respectively for 10min at the N_2 atmosphere to achieve films with about 400-nm-thickness. The resistivity of these thin films was calculated according to the results of the square resistance measured by using the four-probe meter. The crystalline structures of five different ZnO thin films which were pyrolyzed at 250°C, 300°C, 350°C, 400°C, 450°C, respectively and RTA at 600°C were characterized by X-ray Diffraction (XRD) with Cu Kα radiation.

Besides, the crystalline structures of ZnO thin films on such different substrates as SiO_2, Pt, PZT and BFO were characterized by XRD.

Four different ferroelectric-semiconductor heterojunctions namely $(Bi_{0.95}La_{0.05})FeO_3$ (BLFO)/ZnO, $Bi(Fe_{0.95}Mn_{0.05})O_3$ (BFMO)/ZnO, $BiFeO_3$(BFO)/ZnO, and PZT/ZnO were fabricated by sol-gel process on $Pt/Ti/SiO_2/Si$ substrate. Their electrical properties were measured and the mechanisms and the influencing factors were investigated. The drain current versus gate voltage transfer characteristics of the back-gated FETs in this work was also analyzed and investigated. The stoichiometric BFO, BLFO, and BFMO solutions (Toshima

MFG, Ltd.) of 0.2mol/Kg were spin-coated on Pt/Ti/SiO$_2$/Si substrate at 3000rpm for 30s. After each coating cycle, it was dried on a hotplate at 240°C for 3min, and then prefired at 350°C for 8min in air. This process was repeated several times to obtain approximately 400nm thick films. The as-deposited thin films were introduced to a RTA process at 550°C for 10min at the N$_2$ atmosphere, which has been proved effective to suppress the secondary phases and modify the BFO material. The PZT solution was spin-coated at 3000rpm for 25s. Then the wet films were dried at 350°C for 5min. The thickness of PZT films would reach to 400nm by repeating this process for several times. Then the sample was annealed at 650°C for 4min in O$_2$ atmosphere.

The current-voltage (I-V) characteristics of these four stack structures were measured by Agilent 4200 system with bottom electrodes grounded and top electrodes applied to a sweeping voltage of +10 V→-10 V→+10 V.

DISCUSSION

The influence of pyrolysis temperature and annealing temperature on the resistivity and the crystal morphology of ZnO thin films are studied and the results are shown in Figure 2. Figure 2(a) exhibits the resistivity properties of ZnO thin film with different pyrolysis temperatures under each certain annealing temperature, in which the minimum resistivity is achieved at the pyrolysis temperature of 400°C for each annealing temperature respectively in this experiment. Moreover, among the annealing temperatures in this experiment, the lowest resistivity is always achieved at the annealing temperatures of 600°C under the same pyrolysis temperature. Figure 2(b) exhibits that when annealing temperature is at 600 °C, the pyrolysis temperature has a little effect on orientation to crystallize of ZnO thin films. It is found that ZnO thin films show a strongest peak in (002) orientation and the intensity in (101) orientation increases a little with increasing pyrolysis temperature.

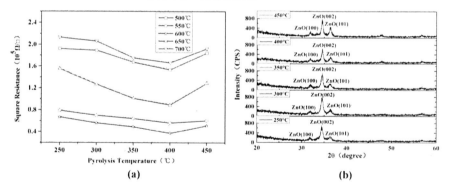

Figure 2. (a) Resistivity of different ZnO thin films at diffident pyrolysis or annealing temperature (b) XRD of different ZnO thin films at diffident pyrolysis temperature

The XRD patterns of ZnO thin films on diffident substrate materials are showed in Figure 3. It is observed that diffident substrate materials have little effect on orientation to crystallize of

ZnO thin films, while all the ZnO thin films reveal three major orientations: (100), (002) and (101) and show a strongest peak in (002) orientation. Moreover, different substrate materials have significant impacts on the crystallization intensities of the ZnO thin films. Among these different specimens, ZnO thin film has the strongest crystallization intensity on Pt substrate and the weakest crystallization intensity on SiO_2 substrate.

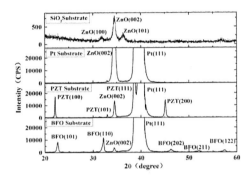

Figure 3. XRD patterns of ZnO thin films on diffident substrate materials

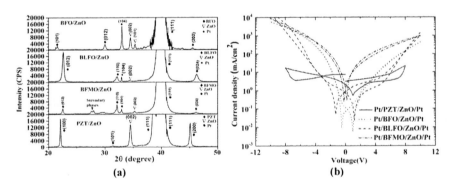

Figure 4. (a) XRD of BFO/ZnO, BLFO/ZnO, BFMO/ZnO and PZT/ZnO structures (b) I-V characteristics of BFO/ZnO, BLFO/ZnO, BFMO/ZnO and PZT/ZnO structures

XRD patterns of BFO/ZnO, BLFO/ZnO, BFMO/ZnO, and PZT/ZnO structures are shown in Figure 4(a). It can be seen that all the diffraction peaks can be indexed, while no non-perovskite phases such as $Bi_2Fe_4O_9$ are observed. The secondary phases appear in BFMO/ZnO structure, in which an additional secondary peak appearing at 2θ ~ 28° could be indexed as the diffraction peak of Bi_2O_3. ZnO intends to be (002) oriented in BLFO/ZnO and PZT/ZnO and

(101) oriented in BFMO/ZnO, while ZnO reveals two orientations that are (002) and (101) in BFO/ZnO. Polycrystalline BFO, BLFO and BFMO thin films in our experiments exhibit mixed peaks such as (012), (110), (104), and (024) with some secondary phases that are Bi_2O_3 as discussed above, while PZT exhibits peaks such as (100), (101), (200). Emerging and shifting of the diffraction peaks of ferroelectric layer can be observed, indicating a structural transformation of the ferroelectric layer induced by La or Mn doping. The (012) peak exhibits relatively larger intensity and moves to a smaller angle in BLFO/ZnO compared to that in BFO/ZnO, suggesting a structural transformation and a crystallization enhancement in a preferred orientation induced by La substitution. Moreover, (104) phase appears and the intensity of (110) phase decreases in BLFO/ZnO and BFMO/ZnO, compared with that in BFO/ZnO. These variations mentioned above induced by impurity atom substitution can also conduce to suppress the leakage properties of BFO thin film and enhance the performance of BFO/ZnO heterojunction.

Figure 4(b) plots the current-voltage characteristics of BFO/ZnO, BLFO/ZnO, BFMO/ZnO, and PZT/ZnO structures. As shown in this figure, the bi-layered structures of BFO/ZnO, BLFO/ZnO and BFMO/ZnO exhibit asymmetric I-V characteristics include a higher resistance at positive voltage and a lower resistance at negative voltage. The dominating factors that lead to such behaviors for BFO/ZnO, BLFO/ZnO and BFMO/ZnO structures are ohmic conduction, grain boundary limited behavior and space-charge-limited current (SCLC) behavior in low electric fields and Fowler-Nordheim tunneling in high electric fields, respectively [7]. Compared with BFO(BLFO,BFMO)/ZnO heterostructures, the leakage current density in PZT/ZnO heterostructure is relatively stable which also can be maintained at a low range.

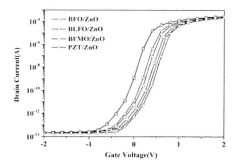

Figure 5. Transfer characteristic curves of MSIM-structure FETs based on different materials

Finally, these four different MSIM-structure back-gated ferroelectric FETs utilizing BLFO/ZnO, BFMO/ZnO, BFO/ZnO and PZT/ZnO, respectively were investigated based on Silvaco. The transfer characteristics of drain current versus gate voltage are shown in Figure 5. The drain voltage is kept in a fixed value and the gate voltage is scanned between ±2V, and typical hysteresis loops due to the ferroelectricity are obtained. The obvious memory windows can be observed in the MSIM structures under the modulation effect of remnant polarization on the channel resistance, including counterclockwise loop and a drain current switching ratio up to two orders of magnitude. Among them, the largest memory window is obtained in the FET with BLFO/ZnO structure, while the smallest is obtained in that with BFO/ZnO structure.

CONCLUSIONS

In summary, different ferroelectric thin films, such as PZT, BFO, BLFO and BFMO, are investigated for their influences on related heterostructures with ZnO and MSIM structures utilized in back-gated ferroelectric FETs. Firstly, different ZnO thin films fabricated by sol-gel process under different pyrolysis temperature, annealing temperature or substrate materials are studied, in which the most ideal pyrolysis and annealing temperature is 400°C and 600°C, respectively, and the substrates of PZT and BFO can promote the crystallization in orientation (002) of ZnO. Secondly, the alterations in preferred crystallization orientations appear in BLFO/ZnO and BFMO/ZnO conduce to find a better impurity doping method to suppress the leakage properties of BFO thin film and enhance the performance of BFO/ZnO heterostructures. Then, the bi-layered structures of BFO/ZnO, BLFO/ZnO and BFMO/ZnO exhibit asymmetric I-V characteristics include a low resistance at positive voltage and a high resistance at negative voltage, while the leakage current density in PZT/ZnO heterostructure is relatively stable which also can be maintained at a low range. Moreover, the transfer characteristics of MSIM-structure FETs based on PZT/ZnO, BFO/ZnO, BLFO/ZnO and BFMO/ZnO are investigated based on Silvaco. Obvious memory windows are observed in the MSIM structures under the modulation effect of remnant polarization on the channel resistance, which also indicates the positive influences of impurity atom substitution in BFO thin film for the MSIM-structure FETs.

ACKNOWLEDGEMENTS

The authors thank the support from the Central University Research Funds of China under No. ZYGX2013J039.

REFERENCES

1. D.D. Lu, M.V. Dunga, C.H. Lin, A.M. Niknejad and C.M. Hu, *Solid-State Electron.* **62** (1), 31 (2011).
2. I.Y. Yang, C. Vieri, A. Chandrakasan and D.A. Antoniadis, *IEEE Trans. Electron Devices* **44** (5), 822 (1997).
3. H.P. Chen, V.C. Lee, A. Ohoka, J. Xiang and Y. Taur, *IEEE Trans. Electron Devices* **58** (8), 2401 (2011).
4. W.Y. Fu, Z. Xu, X.D. Bai, C.Z. Gu and E. Wang, *Nano Lett.* **9** (3), 921 (2009).
5. Z. Jia, M.M. Zhang and T.L. Ren, *Ferroelectrics.* **421** (1), 92 (2011).
6. N.W. Zhang, Z. Jia, M.M. Zhang and T.L. Ren, *Mater. Res. Soc. Proc.* **1368** (1), (2011).
7. J.L. Xu, Z. Jia, N.W. Zhang and T.L. Ren, *J. Appl. Phys.* **111** (7), 074101 (2012).
8. Y. Ryu, T.S. Lee, J.A. Lubguban, H.W. White, B.J. Kim, Y.S. Park and C.J. Youn, *Appl. Phys. Lett.* **88** (24), 241108 (2006).
9. I.D. Kim, Y.W. Choi and H.L. Tuller, *Appl. Phys. Lett.* **87** (4), 043509 (2005).

10. L. Pintilie, C. Dragoi, R. Radu, A. Costinoaia, V. Stancu and I. Pintilie, *Appl. Phys. Lett.* **96** (1), 012903 (2010).
11. Y. Kato, Y. Kaneko, H. Tanaka and Y. Shimada, *Jpn. J. Appl. Phys.* **47**, 2719 (2008).
12. Z. Jia, T.L. Ren, Z.G. Zhang, T.Z. Liu, X.Y. Wen, H. Hu and L.T. Liu, *J. Phys. D: Appl. Phys.* **39** (12), 2587 (2006).
13. M.M. Zhang, Z. Jia and T.L. Ren, *Solid-State Electron.* **53** (5), 473 (2009).
14. Z. Jia, T.L. Ren, T.Z. Liu, H.U. Hong, Z.G. Zhang, D. Xie and L.T. Liu, *Chin. Phys. Lett.* **23** (4), 1042 (2006).
15. T. Choi, S. Lee, Y.J. Choi, V. Kiryukhin and S.W. Cheong, *Science.* **324** (5923), 63 (2009).
16. G. Catalan and J.F. Scott, *Adv. Mater.* **21** (24), 2463 (2009).
17. H. Uchida, R. Ueno, H. Funakubo and S. Koda, *J. Appl. Phys.* **100** (1), 014106 (2006).
18. K.Y. Yun, M. Noda and M. Okuyama, *J. Korean Phys. Soc.* **42**, 1153 (2003).
19. V.R. Palkar, J. John and R. Pinto, *Appl. Phys. Lett.* **80** (9), 1628 (2002).
20. Y.P. Wang, L. Zhou, M.F. Zhang, X.Y. Chen, J.M. Liu and Z.G. Liu, *Appl. Phys. Lett.* **84**, 1731 (2004).
21. X.D. Qi, J. Dho, R. Tomov, M.G. Blamire and J.L. MacManus-Driscoll, *Appl. Phys. Lett.* **86** (6), 062903 (2005).

Mater. Res. Soc. Symp. Proc. Vol. 1633 © 2014 Materials Research Society
DOI: 10.1557/opl.2014.118

Highly reliable passivation layer for a-InGaZnO thin-film transistors fabricated using polysilsesquioxane

Juan Paolo Bermundo[1], Yasuaki Ishikawa[1], Haruka Yamazaki[1], Toshiaki Nonaka[2], and Yukiharu Uraoka[1]

[1]Graduate School of Materials Science, Nara Institute of Science and Technology, 8916-5 Takayama-cho, Ikoma, Nara 630-0192, Japan

[2]AZ Electronic Materials Manufacturing Japan K.K. 3330 Chihama, Kakegawa-shi, Shizuoka, 437-1412, Japan

ABSTRACT

Polysilsesquioxane passivation layers were used to passivate bottom gate a-InGaZnO (a-IGZO) thin film transistors (TFT). The a-IGZO TFTs passivated with polysilsesquioxane showed highly stable behavior during positive bias stress, negative bias stress, and negative bias illumination stress. A voltage threshold shift of up to 0.1 V, less than -0.1 V and -2.3 V for positive bias stress, negative bias stress, and negative bias illumination stress, respectively. We also report the effect of reactive ion etching (RIE) on the electrical characteristics of a-InGaZnO (a-IGZO) thin-film transistors (TFT) passivated with the polysilsesquioxane-based passivation layers. We show how post-annealing treatment using two different atmosphere conditions: under O_2 ambient and combination of N_2 and O_2 ambient (20% O_2), can be performed to recover the initial characteristics. Furthermore, we present a highly stable novel polysilsesquioxane photosensitive passivation material that can be used to completely circumvent the reactive ion etching effects.

INTRODUCTION

Amorphous InGaZnO (a-IGZO) has been extensively studied as replacement for a-Si channel materials in thin film transistors [1-3]. This is not only due to its higher mobility, lower threshold voltage (V_{th}) and much lower off current compared to a-Si but also because of other remarkable properties such as low temperature fabrication, good uniformity and transparency. This makes a-IGZO attractive in current display applications such as organic light emitting diode displays [1,4] and in future display devices where flexibility and transparency are necessary [5].

However, despite having impressive properties, a-IGZO TFTs especially those having a bottom gate structure suffer from instability because the backchannel is exposed to ambient effects. The influence of these ambient effects such as moisture, adsorbed oxygen and post fabrication damage have been extensively discussed [6-9]. The proposed solution to this problem has been to either adopt a different sample structure, for example: top gate structure [2] and etch stop structure [10] or to coat the TFT with a passivation layer. Both inorganic passivation layers (Al_2O_3 [11], SiO_x [12], SiN_x [13]) and organic passivation layers such as CYTOP [14] and photoacryl [9] have been proposed. Nevertheless, there are some disadvantages that arise from these proposed solutions. For instance, using a different structure such as top gate structure requires additional mask and patterning process steps. Similarly, inorganic passivation layers require much more complicated vacuum processes. Although organic passivation layers can be fabricated more easily compared to their inorganic counterparts, most that have been reported have large V_{th} shifts (ΔV_{th}) during bias stress. We have recently reported an organic passivation layer based on polysilsesquioxane that is not only easy to fabricate but also highly reliable [15]. In this work, we will show the effect of reactive ion etching (RIE) on a-IGZO TFTs passivated with polysilsesquioxane and that these TFTs have high stability after post annealing. We will also show that a-IGZO TFTs coated with a photosensitive passivation layer based on polysilsesquioxane have high stability.

EXPERIMENT

Fabricated a-IGZO TFTs had a bottom gate top contact structure. A 70 nm thick a-IGZO layer was deposited on n-doped Si substrate (resistivity<0.002Ω·cm) with a 100 nm thermally oxidized SiO_2 layer by RF magnetron sputtering deposition at room temperature. Si and SiO_2 are the gate and gate insulator, respectively. The source and drain electrodes were a stack of 70 nm Mo and 20 nm Pt deposited via RF magnetron sputtering deposition and patterned using the lift-off technique. Samples were then subjected to post annealing treatment at a temperature of 300 °C for 2 hours in N_2/O_2 ambient atmosphere.

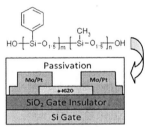

Figure 1. Polysilsesquioxane and passivated a-IGZO bottom gate structure.

Fabricated a-IGZO TFTs were then passivated using two different passivation materials based on polysilsesquioxane. Figure 1 illustrates the polysilsesquioxane structure and the resulting passivated a-IGZO TFT structure. Both passivation materials have similar general structure and only differ on the type of constituent group attached on the Si-O backbone of polysilsesquioxane. For instance, sample Me 100 is passivated with polymethylsilsesquioxane which only has methyl groups as constituent groups. On the other hand, sample Me 60/Ph 40 passivated by a copolymer of methylsilsesquioxane and phenylsilsesquioxane which had methyl and phenyl groups with a ratio of 3:2. In general, phenylsilsesquioxane is copolymerized with methylsilsesquioxane to address the brittleness of methylsilsesquioxane [16]. More details about the polysilsesquioxane passivation layers used in this work is reported in Ref. [15].

The passivation layers were coated using a simple solution process. The passivation layer was initially spin-coated on the a-IGZO TFT at a main spin of 3000 rpm for 15 s. A 2-step heating process was then performed: first, prebaking at 130 °C for 90 s, then post-baking in air at 300 °C for 1 hour. Contact holes were formed by dry etching using inductively coupled plasma reactive ion etching (ICP-RIE) process using a $CF_4/O_2/Ar$ gas mixture. Transfer characteristics were measured after the dry etching. Two different post annealing treatments were then performed by changing the oxygen ratio in the ambient atmosphere. A set of samples were annealed for 2 h at 300 °C under O_2 atmosphere, while another set of samples were annealed using the same time and temperature conditions under a mixed N_2/O_2 atmosphere (20% oxygen). Transfer characteristics and the reliability to bias stress was tested for 10000 s after the post annealing treatment by applying a gate voltage (V_g) of 20 V for positive bias stress (PBS), -20 V for negative bias stress (NBS), and -20V plus illumination with halogen light for negative bias illumination stress (NBIS). TFTs had a channel width and length of 90 and 10 μm, respectively.

Another set of a-IGZO TFTs were also coated with a photosensitive passivation layer by using a simple solution process. The photosensitive passivation is different from its non-photosensitive counterpart by the addition of an SiO_2 constituent group. Also, contact holes were formed by UV photolithography instead of RIE. Like the non- photosensitive passivation layers, a 2 step heating process was used to coat photosensitive polysilsesquioxane passivation: prebake at a lower temperature of 100 °C for 90 s and post bake in N_2 atmosphere for 2 hours also at a lower temperature of 250 °C. The passivated TFTs were also post annealed in O_2 at 250 °C.

DISCUSSION

Transfer characteristics after ICP-RIE show a large change in the electrical characteristics of both Me 100 and Me 60/ Ph 40 samples as shown in figures 2(a-d). Aside from the clear large negative shift in V_{th}, there is also a noticeable increase in the on-current and mobility (data not shown) after ICP-RIE. Furthermore, degradation in the subthreshold swing (S) value was also observed. Yuan $et.\ al.$ reported the effect of O_2 plasma on hydrogensilsesquioxane (HSQ) – a material similar to the passivation used in this case [17]. They reported that O_2 plasma treatment damaged the HSQ films and that an annealing treatment in ambient N_2 can be performed to reverse the damage [10]. In this study, a post-annealing treatment was also performed to recover the initial characteristics. Figures 2(a-d) also illustrate the recovery of electrical characteristics after both post-annealing treatment. Nevertheless, post-annealing under O_2 atmosphere is better for the complete recovery of electrical properties such as V_{th} as shown in the comparison between fig 2e and 2f. Even if the ΔV_{th} after RIE for the TFTs to be O_2 annealed were very large, V_{th} recovery is achieved after O_2 post annealing. Nonetheless, TFTs annealed with either post annealing condition showed similar behavior after RIE and post annealing. Good uniformity in their V_{th} can also be observed which shows the effectiveness of the post-annealing process in controlling and recovering the V_{th}.

Bias Stress Reliability

The difference between the post-annealing conditions is more pronounced when bias stress reliability is tested. We show that post annealing in O_2 is especially effective in improving the reliability of passivated a-IGZO TFTs. Table 1 summarizes the ΔV_{th} after PBS, NBS, and NBIS for both post annealing conditions. It is evident that for both passivated samples, post annealing in O_2 yielded smaller ΔV_{th}. We also observed that hump effect is less pronounced for both O_2

annealed samples than their N_2/O_2 annealed counterparts. The O_2 annealed Me 60/ Ph 40 sample showed very good reliability during PBS with only a ΔV_{th} of 0.1 V compared to 0.29 V for the N_2/O_2 case. A similar trend can be observed for NBS where the ΔV_{th} is smaller for the O_2 post-annealed samples. The effect of O_2 post-annealing on the reliability is especially highlighted in the case of Me 100 during NBIS. Only the N_2/O_2 annealed Me 100 had a very large ΔV_{th} (see Fig. 3) during NBIS. All other samples, especially those that were O_2 annealed had smaller ΔV_{th}.

Table I. Comparison of ΔV_{th} after bias stress for Me 100 and Me 60 Ph/40 samples that were either post annealed in O_2 or N_2/O_2 ambient atmosphere.

	Me 100 ΔV_{th} (V)		Me 60/ Ph40 ΔV_{th} (V)	
	N_2/O_2 Ambient	O_2 Ambient [15]	N_2/O_2 Ambient	O_2 Ambient [15]
Positive Bias Stress	0.42	0.49	0.29	0.10
Negative Bias Stress	-0.23	-0.08	-0.84	-0.09
Negative Bias Illumination Stress	-11.38	-2.45	-2.42	-2.34

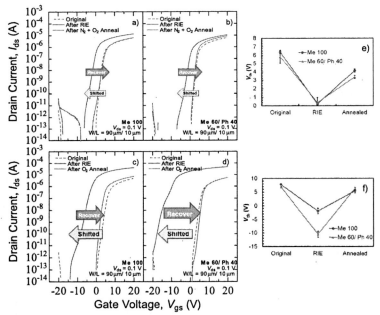

Figure 2. Transfer characteristics (V_{ds} = 0.1 V) of a) Me 100 and b) Me 60/ Ph 40 after ICP-RIE and post annealing under N_2/O_2 atmosphere (20% O_2). Another set of transfer curves after ICP-RIE and O_2 annealing are shown for c) Me 100 and d) Me 60/Ph 40. The V_{th} of Me 100 and Me 60/Ph 40 after RIE and post annealing under e) N_2/O_2 and f) O_2 ambient are also shown

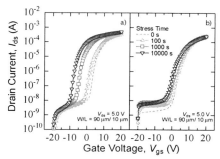

Figure 3. Variation of transfer curves of N_2/O_2 post annealed a) Me 100 and b) Me 60/ Ph 40 under NBIS for 10000 s (V_{gs} = -20 V, illumination by halogen lamp, light intensity: ~ 27,000 nit). Variation of transfer curves of the O_2 post annealed case under NBIS can be seen in Ref. 15

NBIS instability is largely caused by oxygen vacancies (V_O) [18]. We have recently reported that a high C and H concentration in the a-IGZO layer can contribute to the higher reliability especially during NBIS [15]. Figure 4 summarizes how H can contribute to the V_O reduction. The H can occupy the V_O site to form H at V_O site (H_O) which then reduces the V_O density. Reducing the amount of V_O can lead to greater stability during NBIS.

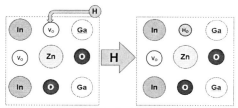

Figure 4. Schematic of H infusion into the V_O site leading to V_O reduction.

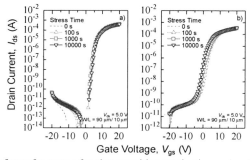

Figure 5. Variation of transfer curves for photosensitive passivation under a) PBS and b) NBIS.

Photosensitive Passivation

The effect of RIE can be completely circumvented by using a photosensitive passivation. In this case, the contact holes are formed by UV photolithography and not RIE. Figure 5 shows the evolution of transfer curves after PBS and NBIS. In this case, a very small ΔV_{th} of 0.43 V after PBS and -2.39 V after NBIS was observed for TFTs passivated with the photosensitive material. This is good considering that the TFTs were post annealed at a lower temperature of 250 °C. The transfer curves also show no hump effect or any degradation in the subthreshold region.

CONCLUSIONS

We have presented novel polysilsesquioxane based passivation layers that are easy to fabricate and are highly reliable. We showed that reactive ion etching can alter the electrical characteristics of a-IGZO TFT and that it can be reversed by post annealing especially under O_2 ambient atmosphere. We also reported a novel photosensitive polysilsesquioxane material that has high reliability. These results show the high potential of polysilsesquioxane as effective passivation materials for a-IGZO TFTs.

REFERENCES

1. T. Kamiya, K. Nomura, and H. Hosono, Sci. Technol. Adv. Mater., **11**, 044305 (2010).
2. K. Nomura, H. Ohta, A. Takagi, T. Kamiya, M. Hirano, and H. Hosono, Nature, **432**, 488 (2004).
3. J. S. Park, W.-J. Maeng, H.-S. Kim, and J.-S. Park, Thin Solid Films, **6**, 1679 (2012).
4. J.-S. Park, T.-W. Kim, D. Stryakhilev, J.-S. Lee. S.-G. An, Y.-S Pyo, D.-B Lee, Y. G. Mo, D.-U. Jin and H. K. Chung, Appl. Phys. Lett., **95**, 013503 (2009)
5. E. Fortunato, P. Barquinha, and R. Martins, Adv. Mater., **24**, 2945 (2012)
6. D. H. Kang, H. Lim, C. J. Kim, I. H. Song, J. C. Park, and Y. S. Park, Appl. Phys. Lett., **90**, 192101 (2007).
7. K.-H. Lee, J. S. Jung, K. S. Son, J. S. Park, T. S. Kim, R. Choi, J. K. Jeong, J.-Y. Kwon, B. Koo, and S. Lee, Appl. Phys. Lett., **95**, 232106 (2009).
8. J.-S. Park, J. K. Jeong, H.-J. Chung, Y.-G. Mo, and H. D. Kim, Appl. Phys. Lett., **92**, 072104 (2008).
9. J. K. Jeong, H. W. Yang, J. H. Jeong, Y.-G. Mo, and H. D. Kim, Appl Phys. Lett., **93**, 123508 (2008).
10. M. Mativenga, J. W. Choi, J. H. Hur, H. J. Kim, and J. Jang, Journ. of Info. Disp., **12**, 47 (2011).
11. C. H. Ahn, K. Senthil, H. K. Cho, S. Y. Lee, Nature Sci. Rep., **3**, 2737 (2013)
12. S.-H. Choi, and M.-K. Han, IEEE Electron Device Lett., **33**, 396 (2012).
13. T. Arai, N. Morosawa, K. Tokunaga, Y. Terai, E. Fukumoto, T. Fujimori, T. Nakayama, T. Yamaguchi and T. Sasaoka, SID Symposium Digest of Tech. Papers, **41**, 1033 (2010).
14. S.-H. Choi, J.-H. Jang, K. Jang-Joo, and M.-K. Han, IEEE Elect. Dev. Lett., **33**, 381 (2012)
15. J. P. Bermundo, Y. ishikawa, H. Yamazaki, T. Nonaka, and Y. Uraoka, ECS Journal of Solid Stat. Sci and Tech. (2013) (in press)
16. R. H. Barney, M. Itoh, A. Sakakibara, and T. Suzuki, Chem. Rev., **95**, 1409 (1995).
17. Q. Yuan, G. Yin, and N. Zhaoyuan: Plasma Sci. and Technol., **15**, 86 (2013).
18. M.D.H Chowdhury, P. Migliorato, and J. Jang, Appl. Phys. Lett., **97**, 173506 (2010).

AUTHOR INDEX

Allen, Martin Ward, 13, 51

Baldwin, D., 87
Berengue, O.M., 25
Bermundo, Juan Paolo, 139

Chen, Bing, 105
Chiquito, A.J., 25
Chuang, Ta-Ko, 95
Cliff, J., 87

Dang, G.T., 13
Dell, J.M., 87
Dietrich, Christof Peter, 51
Durbin, S.M., 13

Faraone, L., 87

Gao, Bin, 105
Goto, Hiroshi, 55
Grundmann, Marius, 51, 101, 123

Hanawa, T., 19
Hatalis, Miltiadis K., 95
Hayashi, Kazushi, 55
Heinhold, Robert, 51
Hering, K.P., 3
Hino, Aya, 55
Hyland, A., 13

Ishikawa, Yasuaki, 139

Jeffery, R., 87
Jia, Ze, 131

Kang, Jinfeng, 105
Katiyar, Ram S., 111
Kato, Takahiro, 61
Kennedy, R.J., 13
Khishigjargal, Tegshjargal, 69
Kikuchi, N., 19
Kim, Hyung-Suk, 51

Klar, Peter J., 81
Kramm, B., 3
Krishnan, R.N., 87
Kugimiya, Toshihiro, 55
Kung, Jerry Ho, 95

Lenzner, J., 123
Lindemuth, Jeffery, 43
Liou, Juin J., 131
Liu, Dong, 105
Liu, Jizhi, 131
Liu, Lifeng, 105
Liu, Xiaoyan, 105
Liu, Zhiwei, 131
Longo, E., 25

Mahmoudabadi, Forough, 95
Mamiya, T., 19
Marshall, Luke G., 43
Martyniuk, M., 87
Medina, G., 13
Medvedeva, Julia E., 37
Meyer, B.K., 3
Misra, Pankaj, 111
Morita, Shinya, 55
Müller, Stefan, 51
Mullins, C. Buddie, 43
Murat, Altynbek, 37

Nagatomi, Eichi, 61
Nakamura, Toshihiro, 117
Nishio, K., 19
Nonaka, Toshiaki, 139

Ohishi, Koichiro, 61

Pereira-da-Silva, Marcelo A., 25
Piper, L.F.J., 13
Polity, A., 3
Pontes, D.S.L., 25
Pontes, F.M, 25
Portz, A., 3

Pratap, Rudra, 75

Reeves, R.J., 13
Reindl, Christian T., 81
Rettie, Alexander J.E., 43

Sakai, Osamu, 117
Sander, Thomas, 81
Schlupp, Peter, 51, 101
Schmidt, Florian, 51, 123
Sharma, Yogesh, 111
Shivashankar, S.A., 75
Silva, K.K.M.B.D., 87
Singh, Nagendra Pratap, 75
Stampe, P.A., 13

Takanashi, Yasuyuki, 55
Tamayama, Yasuhiro, 61
Tao, Hiroaki, 55
Tonooka, K., 19

Ueda, Kazuyoshi, 69
Uraoka, Yukiharu, 139

von Wenckstern, Holger, 51, 101, 123

Wahila, M.J., 13
Wang, R., 19
Wang, Yiran, 105
Woodward, R.C., 87
Wu, Xiao, 131

Xu, Jianlong, 131

Yamada, Masaki, 117
Yamaguchi, Naoya, 61
Yamazaki, Haruka, 139
Yasui, Kanji, 61

Zhang, Mingming, 131
Zhang, Naiwen, 131
Zhang, Z., 123
Zhou, Jianshi, 43

SUBJECT INDEX

amorphous, 101

Bi, 43

catalytic, 61
chemical vapor deposition (CVD) (deposition), 61

deep-level transient spectroscopy (DLTS), 51
defects, 37, 51, 55, 75, 81
devices, 123
dielectric, 111
dopant, 131

electrical properties, 25, 55, 95
electronic material, 43
electronic structure, 37
epitaxy, 19, 25

ferroelectric, 131

hardness, 87

II-VI, 123
ion-beam assisted deposition, 87

laser ablation, 19
luminescence, 75

memory, 105
microelectronics, 95
microstructure, 51
molecular beam epitaxy (MBE), 13

nanostructure, 75

optical properties, 117
oxide, 3, 13, 37, 81, 105, 139

passivation, 139
photoconductivity, 139
photovoltaic, 3
piezoelectric, 19
plasma deposition, 117

Raman spectroscopy, 81, 111

self-assembly, 69
semiconducting, 55
sensor, 123
simulation, 69
Sn, 101
sol-gel, 105, 131
sputtering, 3

thin film, 13, 25, 61, 87, 95, 117
transparent conductor, 69

V, 43

x-ray diffraction (XRD), 111

Zn, 101

CPSIA information can be obtained at www.ICGtesting.com
Printed in the USA
LVOW10*1046020514

384112LV00001B/1/P

9 781605 116105